THE EVOLUTION OF DEATH

SUNY Series in Philosophy and Biology
David Edward Shaner, editor

the evolution of death

why we are living longer

stanley shostak

STATE UNIVERSITY OF NEW YORK PRESS

Published by
State University of New York Press, Albany

For information, address State University of New York Press,
194 Washington Avenue, Suite 305, Albany, NY 12210-2384

Production by Marilyn P. Semerad
Marketing by Susan M. Petrie

Library of Congress Cataloging in Publication Data

Shostak, Stanley.
  The evolution of death : why we are living longer / Stanley Shostak.
      p. ; cm. — (SUNY series in philosophy and biology)
  Includes bibliographical references and index.
  ISBN-13: 978-0-7914-6945-3 (hardcover : alk. paper)
  ISBN-10: 0-7914-6945-X (hardcover : alk. paper)
  ISBN-13: 978-0-7914-6946-0 (pbk. : alk. paper)
  ISBN-10: 0-7914-6946-8 (pbk. : alk. paper) 1. Death. 2. Aging. 3.
Life expectancy.
  [DNLM: 1. Death. 2. Aging. 3. Evolution. 4. Life
Expectancy—trends. WT 116 S559e 2006] I. Title. II. Series.
  QP87.S43 2006
  613.2—dc22
                                              2005037275

10 9 8 7 6 5 4 3 2 1

To A. B. and G. P.

Contents

Illustrations

Preface

The changes in expectation of life from the middle of the seventeenth century to the present time where the records are most extensive and reliable appear to furnish a record of a real evolutionary progression. In this respect at least man has definitely and distinctively changed, as a race, in a period of three and a half centuries.

—Raymond Pearl, *The Biology of Death*

". . . Death is like birth, painful, messy and undignified. Most of the time anyway." She thought, Perhaps it's just as well. Reminds us that we're animals. Maybe we'd do better if we tried to behave more like good animals and less like gods.

—P. D. James, *Death in Holy Orders*

Science has always been my favorite form of whodunit, especially when the scientist discovers a tantalizing mystery and solves it with clever experiments or observations. Regrettably, some very tantalizing mysteries remain on the back shelf of science, never having made it to the bestseller list. Death is one such mystery. We all know that death, like reproduction and metabolism, is a fundamental feature of life—only living things can die—and we also know that death is somehow inherited from generation to generation, but exactly what death is adapted to and how it evolves are mysteries that have remained sub-rosa. Until now!

The Evolution of Death is about to change death from a dead subject into a vital one, burgeoning with those concepts and consequences that traditionally arouse curiosity and command attention about life. The problem is that death, like taxes (to take a page from Benjamin Franklin), is thought to be inevitable and unchanging. Remarkably, while belief in the inevitability of many things, such as war, poverty and crime, has slackened in the last few years, belief in the inevitability of death has remained unshaken. Continents have been seen to move, the appearance and disappearance of oceans has been acknowledged, and even stars have waxed and waned, but the immutability of death continues to ride the storm. Well, no more!

In fact, registries and census data have recorded evidence of *death's evolution* for centuries, but hardly anyone took notice. These data were and are fobbed off as consequences of improvements in lifestyle brought about through agriculture, industrial, technological and informational revolutions, and modern, urban living. But these same data also testify to death's progress toward aptness, downstream in the flow of life. In fact, the mystery solved here is whether the human form of death has been evolving for millennia, inexorably achieving greater refinement, efficiency, and cost effectiveness.

Fundamentally, *death is the process of making corpses* from living things. Hence, *death evolves by making corpses better*—making corpses more easily, more efficiently, and with less disruption to life, which is to say, corpses that waste less of life's precious material than in the past. Death thus feeds back onto life, turning the body into a corpse only after life is exhausted. Because evolution supplies new and improved models of death, life is becoming longer, fuller, and healthier.

The Evolution of Death traces these improvements in death to changes in the life cycle: by lubricating life's cycle, death greases the way to better life. Moreover, life cycles with the best lubrication shape future generations. If the expansion of the human lifespan continues at its present rate or accelerates, as some gerontologists predict,[1] we may very well live indefinitely, and death will truly have died.

But before I abandon the present generation to its untimely fate, allow me to express my profound gratitude for the help I've had writing this book. Truly, if I acknowledged all the sources I have gathered beyond those cited, I would add unconscionably to the length of this preface. Possibly the greatest luxury available to academics is the leisure to wade through the literature until the source of ones "own" idea is found, and the literature on aging and mortality is, if nothing else, one of ideas. I am, indeed, in debt to and in awe of the early aging theorists and can only hope that *The Evolution of Death* is an appropriate tribute to them. Beyond this general acknowledgment, however, I will let them rest in peace.

My more immediate and pressing debts are to the living. Drynda Johnson, head librarian at the Langley Lending Library, University of Pittsburgh, and her assistants, Laura McVey and Ann Rogers, performed yeoman service getting me every book and article I requested, no matter how remote the source or esoteric the subject. Thanks!

I would be shamefully guilty of neglect were I to fail to mention members of Los Angeles' Gerontology Research Group who attended my lecture in February 2003 and offered much needed and appreciated criticism through correspondence. In particular, I am indebted to L. Stephen Coles, M.D., Ph.D. who read Chapter 1 and provided criticism when it was urgently needed, and to Karlis Ullis, M.D. who provided lavish hospitality and profound analysis during my stay in Los Angeles. Thanks again!

I would be remiss were I not to thank my acquisition editor, Jane Bunker of the State University of New York Press, who shepherded the manuscript through many a rough spot. May I add my gratitude to my two anonymous readers who managed to find helpful things to say amidst their criticisms. And may I add my thanks to Robert Olby, my friend and critic, who read and guided me through history in several chapters.

I am also grateful to my son, Daniel Shostak, who worked with me on applying the chronic disease model to life. In fact, it was Dan who made the model work for pies and doughnuts, accordions and bagpipes!

Finally, Marcia Landy, Distinguished Service Professor and film scholar, is my first reader of record and my best and wisest critic. Her innumerable readings of the manuscript for this book, I confess, made it the book that it is today. I could not have had a better, kinder, gentler friend. No amount of thanks is sufficient, but I keep trying.

Introduction:
Death the Mystery

Human beings are near-perfect animals. Of course, we might be improved with a few minor adjustments—strengthening the back for lifting and bearing in an upright posture, broadening hips for ease and safety of childbirth[1]—but, with one major exception, we are exquisitely adapted to our way of life: to survival and reproduction in our terrestrial habitat and agricultural-mechanized-technological ecological niche. Like our mammalian relatives, adult human females provide an excellent womb for gestation and the capacity to nurture offspring with milk of high nutritional quality. Like our primate cousins, human beings have reasonably good color discrimination, stereoscopic vision, sense of balance and acceleration, and adequate olfaction. We also have opposable thumbs, dexterous hands, a bipedal gait, and an immense brain capable of virtually infinite learning and inspired thinking.

Our major flaw is death:[2] We "die like dogs" (or animals generally). Death, which is to say, irreversible damage to the chemistry of life,[3] is the greatest affront to human dignity, especially death accompanying aging. What I have in mind is death due to senescence and decrepitude, the ultimate killers, as opposed to the forms of death that will always be with us—death due to trauma, war, overwork, famine, and infectious disease. Why are we cut off by age, left to decline instead of prosper in our prime? Why are we rendered helpless by senescence? Why are we consigned to intolerable fear, subject to useless pain, and made virtually vegetative in our old age? Indeed, why do "changes occur in a human being from 30 to 70 to increase the chance of dying by roughly 32 fold"?[4]

WHY DO WE DIE?

All these questions smack of mystery, and many mystifying answers have been proposed over the millennia. Some answers, while not incontrovertible and definitive, seem to serve one purpose or another. Answers may take the

1

form of myths and serve a variety of cultural functions, although they do not deal with death materially. Other answers pose scientific explanations and serve the legitimate interests of the science establishment, governmental health agencies, and pharmaceutical, insurance, and health care industries without explaining death fundamentally. One may wonder if "why" is the right question.

First of all, answers to why questions proposed by scientists concerned with the laws of nature and physicians concerned with the nitty-gritty of health care tend to stream off to remote and impersonal causes. Death, on the other hand, is very proximate and personal. Ultimately, answers to fundamental questions should include material causes, and answers to materialist questions must also bear fundamental causes. What is required is a joining of fundamental causes and material causality in a unified principle—finding a thread as well as a needle to stitch answers together into a unified concept of death.

Secondly, answers offered by the man on the street and by religious and philosophical thinkers frequently vanish at the far end of relevance. These answers range from clichés to the esoteric, from slogans on bumper stickers to occult arcana and from sacred doctrine to scientific dogma. No wonder answers have been debated for millennia without resolution!

Casual talk in the barroom, and bedroom about why we die seems broadly unfocused, clouded by confusion, and shaded by mystery. Of course, answers that instill fear or terror of pain or of the beyond may be intended to deter suicide in those in despair or in agony without answering the question. Otherwise, to say "We die because we are born" reveals a primitive fatalism, and to suggest "We die because it's normal" exposes a pessimistic determinism. Such statements may even have the ring of objectivity, scientific reductionism, and statistical predictability, but neither birth nor bell-shaped curves *cause* death in any direct way.

Theologians tell us that we die for any number of reasons, but generally because the human soul or spirit "has a life of its own," an immortal life, as the case may be, and cannot be tied down to a mortal body. As reflected in the thought of great philosophers around the world and throughout the ages, the human soul is considered the nature of being, an active principle of life, of consciousness, conscience, justice, truth, joy, affection, tenderness, cherishing, and love. Thus we die because the departure of this soul deprives us of everything recognizable about human life.

Death is merely the gateway of the immortal soul to eternity. We die because the soul must move on while the body returns to dust. We die to permit the eternal spirit to reach its potential, its emancipation and ascendance. Or we die for less elevated reasons: because one is rewarded or punished for how one has lived ("For the wages of sinne is death"[5]); because God condemned us to die after Adam and Eve sinned in the Garden; and so on.

In addition, the notion that the soul or spirit can survive separation from the body while leaving a corpse behind may relieve the bereaved of anxiety, provide survivors with rationales that calm and cajole, and coerce the despondent back from the abyss to their place in society and to useful employment. But souls have never been seen leaving bodies no matter whose cemetery one visits or how deeply one digs. The question is, can the departure of anything so immaterial as a soul *explain* anything so material as a corpse?

Saying that death is caused by the departure of a living principle from the body would seem more circular than causal. What is more, various bodies, such as those of animals, may live without souls or spirits, although one may reserve the possibility that pets and some domesticated animals have souls. Indeed, even some human beings are said to exist without a soul or spirit—at least as testified by fairy tales and mythologies of wood nymphs and water goblins, to say nothing about horror and fantasy novels, grade B movies, and comic books featuring zombies, vampires, and the "living dead." Such sprites may not be truly human, however, since they are incapable of the human passion for truth, if not for beauty.

But let us bend over backwards to make the case for the departure of a soul as the material cause of death. Let us imagine that the Grim Reaper, Angel of Death, Avenging Angel, or Winged Chariot is *really* an effector of the material transformation of body to corpse. Thus, as the effector liberates or takes possession of the soul, the body turns into a purely material corpse on the verge of returning to dust.[6] Is such an effector a legitimate or an unnecessary hypothesis?

Much of what has been learned through the scientific study of death suggests that death is amply correlated with material causes, but nothing whatsoever would seem to prevent a transcendent effector from using these same causes while turning an organism into a corpse, and one is loath to tell a transcendent effector how to operate. Thus, the issue here is not one of mechanism—how the power over life and death is exercised—but where that power rests. On the one hand, material causality operates blindly, while, on the other hand, a transcendent effector has discretion. Under the aegis of material causes, the same conditions bring about the same end, while decisions over life and death by a transcendental effector could be cogent and might even be subject to approval by higher authority.

The burden for demonstrating a role for a transcendent effector in life and death decisions, therefore, would seem to rest on how much discretion is at the effector's disposal. Indeed, prophets, eager to demonstrate God's mercy, have claimed to have intervened successfully on behalf of those threatened with death, and much religious ceremony, rite, and ritual is devoted to repealing God's fatal sentence. Certainly, the God "that didst dye for me"[7] epitomizes the possibility of discretionary power over life and death.

But can one parry the effector's fatal thrust? Can I truly "lay me down and dee" for bonnie Annie Lawrie?[8] Can soldiers die in the place of their leaders; can martyrs die in the place of their followers? Or, to reverse the field, can victims and scapegoats effectively condemn their persecutors to death? For the most part, substitution does not seem an available option no matter how fervent the appeal or justified the claim. Death does not seem to be the cipher, and a transcendent effector does not seem to be the decipherer. But other possibilities remain.

Another point at which transcendence might work causally occurs when deities or gods decide to interfere in the affairs of mortals. We are told, for example, "As Flies to wanton Boyes are we to th' Gods, / They kill vs for their sport."[9] These interlopers are higher powers that cannot be questioned or held accountable, however, and, thus we cannot hope to learn anything about causality from them.

Philosophers may be more sympathetic than theologians to the desire to explain death materially, but philosophers are not necessarily more successful. Jacques Derrida, the late deconstructionist philosopher of being, suggests that death, like history, must remain a mystery, even in light of the excess of knowledge and detail presently surrounding it. "Philosophy . . . is nothing other than this vigil over death that watches out for death and watches over death, as if over the very life of the soul."[10] For Derrida, the question is not one of adaptation but whether mortal human beings can acknowledge death. "[T]o have the experience of one's absolute singularity and apprehend one's own death, amounts to the same thing."[11] But "singularity" can be the importance, significance or meaning of death without explaining it.

Thomas Nagel, the philosopher of problems, intuition, and discord, on the other hand, has acknowledged death without acknowledging singularity. Building on Epicurus' notion of death as the end of sensation and awareness, Nagel extends "'death' . . . to mean *permanent* death, unsupplemented by any form of conscious survival."[12] Death is thus the making of corpses, turning a living thing into something "dead as a doornail," and that would seem to be as far as philosophical discourse can go. Thus, materiality and causality cannot be integrated in the philosopher's calculus but the dead can be specified.

The picture emerging here is beginning to be complex and detailed enough—the haystack is taking shape—but what is the thread and what is the needle? Let me hint broadly: is the thread life and is the needle evolution?

Part I

How Biology Makes Sense of Death

In the past, death posed a conundrum for biologists: death as such did not seem to perform a function in life, yet death seemed a part of life, since only living things died. Indeed, death did not seem to be one of life's qualities, even though, with few exceptions, it was the end of life. Likewise, death seemed incapable of evolving, since it did not contribute to the fitness of the individual, and genes would not, therefore, determine death.

Part 1 reexamines these premises. If death is part of life, it must take part in life, but how? Chapter 1 emphasizes evolution as the principle that unifies death with life while dismissing some of the false clues that have misled scientists in their quest to make sense of death. Chapter 2 cuts to the chase: if death is part of life, then death must evolve, and, indeed, it does! The chapter goes on to use a chronic disease model of survival to examine two possibilities for lifetime expansion: the accordion and bagpipe models. Only the extension of life's juvenile stage makes sense when tested. Chapters 3 and 4 raise the stakes, finding death's place in the context of life's complexity, lifecycles (chapter 3), and life's far-from-equilibrium thermodynamics and statistical dynamics (chapter 4). Lifecycles connect life to life, while death smoothes the flow between them and lengthens the life span through evolution. Death is not the waste of life it sometimes seems, but the pathway through which energy flows through virtual life and through its fractals.

Chapter 1

Evolution: Death's Unifying Principle

We should also recall, as if we needed reminding, that we are mortal and limited, and thus should remember that the old myths of unrestricted curiosity and the corruption of power are not necessarily fables.
—Simon Conway Morris, *Life's Solution: Inevitable Humans in a Lonely Universe*

Normality seems to have nothing to do with it, for the fact that we will all inevitably die in a few score years cannot by itself imply that it would not be good to live longer.
—Thomas Nagel, *Mortal Questions*

The machine, mon ami, wears out. One cannot, alas, install the new engine and continue to run as before like a motor car.
—Agatha Christie, *Curtain: Hercule Poirot's Last and Greatest Case*

All living things have their own ways of dying or not. I describe these ways in the appendix, but *The Evolution of Death* is primarily concerned with death in *Homo sapiens*—our death. If we are ever to understand death, it will be because we see it as part of life—as evolving. Science got it wrong several times in the past, but the consequences of death's resuscitation, its reinstallation in life, for culture and civilization will be enormous.

DEATH EVOLVES!

In the last few hundred years, human beings have created an environment in which death has been delayed as a result of all sorts of improvements:

sanitation, nutrition, medicine, and so on. Those who most profited from these changes have lived to tell the tale. And their survival and reproduction has shaped the evolution of our death. Consequently, individuals remain young longer and delay aging to their later years. Indeed, so-called natural or age-dependent human death now comes later than at any time in the past.

One struggles vainly to isolate a single cause of death's evolution. For example, levels of dietary sodium and genes both influence each of the age-related biological measures of declining cardiac function, including heart rate, blood pressure, and arterial stiffness. Effects of environmental and genetic factors on aging, dying, and death may be indistinguishable, and particular environments seem to produce phenocopies (that is, environmentally induced mimics of mutations). For example, in model systems, the effects of caloric restriction on enhancing longevity are identical to single gene mutations that increase life span from 30 percent to a doubling or more.[1] The environmental effect set off by reducing the number of calories in the diet converges with the effect of genes encoding members of the insulin-like glucose-metabolism pathway. Like life, death is a facet of underlying continuity, endlessly moving and evolving.

The scale of death's recent evolution is also difficult to grasp, and accepting it may require a thorough reorientation toward life. Instead of imagining death as the antithesis of life, death must be appreciated as an evolving part of life and an adaptation to life. Life must also be seen differently, namely, as incorporating the various aspects of death, such as exchange, feedback, turnover, and regulation. Indeed, death's major features, it turns out, create life as we know it, and even make life possible!

One might think, naively, isn't it ironic that death has evolved toward the accumulation of resources, the prolongation of youth, and the extension of life in succeeding generations? But the irony disappears upon reflection. When we die of old age, it is not because we have failed prematurely to utilize our inborn resources. Those resources—in particular, our stem cells—are invested throughout our lifetime. We die because these resources are exhausted. We die because hardly anything remains (for example, of our stem-cell populations) capable of supporting further life. But the downstream movement of death is a direct consequence of our upstream addition of resources that prolong youthfulness and hence life. In the future, as long as we continue to shape our ecological niche toward longevity, human beings will be born with greater and greater resources and hence increased longevity. It is widely acknowledged that human beings are generally living longer today than ever before, but death will continue to optimize, and as it approaches its apotheosis, death will all but disappear!

Chance, of course, also enters the equation of life,[2] in the sense of reactions that are probabilistic as opposed to deterministic, and constraints on intrinsically stochastic fluctuation and feedback rather than mere alternate pathways and unspecified ranges of variation. Hence chance, along with the environment and genes, enters equations for the accumulation and availability of resources, accounting for the variability of life span.

Thus, death is a part of life. Death evolves when living things accumulate resources, when genes and other hereditary influences provide the pathways that make those resources available, the environment makes them accessible, and chance decides whether or not a resource will be there when needed. Death is subject to natural selection, changing over generations under the auspices of contingency and opportunity. By coming later in life, after the exhaustion of resources, death exhibits the exquisite integration of structure and function peculiar to life. And, hence, death is adaptive. Through its evolution, death increases fitness, emerging from and enhancing reproduction, like other aspects of life. Indeed, we still die, but evolution has made death operate more efficiently and economically than at any time in the past—and death is still evolving.

FALSE CLUES: WHERE SCIENCE GOT IT WRONG

Scientists function to provide worldly solutions to problems and favor numbers and equations over mere words. And scientists are supposed to be sufficiently disinterested when it comes to death to perform their function.

The Nobel Prize–winning zoologist/immunologist and author, Peter Medawar, for example, had no truck with terms pirated from the vernacular, insisting instead on a working understanding. From his vantage point, the terms "life" and "death" "used in scientific contexts [were] far removed from those [contexts] that might arise in common speech . . . [such as] whether the condition of the possible [organ] donor is reversible or not."[3] But even scientists willing to take on eternal verities frame aging, dying, and death within a canonical mold: we die because living things have always died.[4] Thus, we die at the behest of statistics, of a species' finite life span, of killer genes, killer environments, or entropy and the laws of thermodynamics. But do we?

Chapter 1 examines the objectivity of these scientific truths. Several questions are raised in the form of "Do we die at the whim (command, behest) of . . . ?" But to all these questions, the answer is resolutely no. The rejection of these "objective" possibilities ultimately places death on its one firm basis, namely, life.

DO WE DIE AT THE WHIM OF STATISTICS?

Thomas Robert Malthus (1766–1834) should be credited with making an early effort to put a scientific face on the statistics of death. His 1798 *An Essay on the Principle of Population* (largely a polemic on the necessity for appropriation and uneven distribution of wealth, a diatribe against Mr. Pitt's Poor Laws, the parish system, and enclosure of the commons, and a mocking critique of notions of physical immortality) argued "that the power of population is indefinitely greater than the power in the earth to produce subsistence for man," and "in no state that we have yet known has the power of population been left to exert itself with perfect freedom."[5] Therefore, populations are held in check, frequently, but not necessarily, at their subsistence level. According to Malthus, human populations are constrained both positively (preventively), for example, by marriage, virtue, and other moral constraints, and negatively (destructively), for example, by contraception, abortion ("improper arts to conceal"[6]), and premature death. Specifically, the "lower classes . . . suffer from the want of proper and sufficient food, from hard labour and unwholesome habitations . . . [to which] may be added vicious customs with respect to women, great cities, unwholesome manufactures, luxury, pestilence, and war."[7] Later, in *A Summary View of the Principle of Population,* Malthus added to the list of negatives the "whole train of common diseases and epidemics . . . infanticide, plague and famine."[8]

Charles Robert Darwin (1809–1882) "happened to read for amusement Malthus on *Population,* and being well prepared to appreciate the struggle for existence which everywhere goes on . . . [was] at once struck . . . that under these circumstances favourable variation would tend to be preserved, and unfavourable ones to be destroyed."[9] Alfred Russel Wallace (1823–1913), the "other" discoverer of natural selection, admits to a similar "coincidence."[10] But candor aside, Darwin and Wallace were compelled to acknowledge their debt to Malthus if only because his pamphlet was widely read. His doctrine might also have been broadly accepted in Britain, if not elsewhere, as Daniel Todes points out: "[I]t would not be surprising if Darwin's contemporaries, especially those outside of the British cultural context, associated his struggle for existence with specifically British, bourgeois, or Malthusian values."[11]

Of course, Darwin and Wallace were less interested in what kept populations in check than in what unleashed the origin of new species. Thus, Darwinism took Malthus's notion of negative checks onestep further, implying that some organisms were selectively squeezed out or killed while others survived because of their advantageous morphology. Pasted together, Malthusian constraints and Darwinian selection became, in essence, a theory of death creating room at the top, or space for the evolution of improved species. But is this synthesis incontrovertible?

Were death to serve the evolutionary function of creating wiggle room for favorable variants, aging and dying would be especially advantageous in species confronting complex and changing environments simply because the survival of these species might depend on variant organisms that happen to be better adapted to new circumstance than run-of-the-mill organisms. Indeed, sexual reproduction itself seems specialized for producing new varieties of organisms, since sex promotes the mixing of genes as a result of (1) recombination between homologous chromosomes, (2) reshuffling originally maternal and paternal chromosomes during the formation of sex or germ cells, and (3) randomly combining germ cells during fertilization. But reshuffling is at least as likely to destroy favorable combinations of genes as to promote fitness interactions, and the results of recombination in the HIV-1 retrovirus, where recombination is frequent, "challenge hypotheses about the evolution of recombination,"[12] suggesting instead that recombination (or template switching) functions in the repair of single strand breaks. Recombination, thus, may be a consequence of and not the cause of evolution.

A second problem is that Darwinian evolution by natural selection requires a reproductive advantage for the individual being selected, and genes promoting the death of the individual would not seem to promote the individual's reproduction, especially if death came before reproduction! Even selfish genes do not bite the hand that passes them to the next generation.

Ultimately, the notion of death offering an advantage founders on the rocks of the fossil record. In fact, there isn't any—or perhaps just very little—room at the top! While human history may well be a tale written by victors, evolutionary scenarios deciphered from the fossil record are tales written by surviving remnants—castaways, outcasts, refugees, and emigrants—left in the wake of cataclysms or isolating processes.

It is not death, after all, that makes way for variation, but changed ecological circumstance that gives existing variants a chance to emerge. Historically, the species that has been most successful in one era (has cornered the market or found an evolutionarily stable strategy) is a dead end in the next era. Such species are more likely to be too specialized to adapt to new circumstance, even with all their variants thrown into the mix. On the other hand, a peripheral and generalized species, possibly highly dispersed as well, is the one likely to evolve and give rise to new species when the environment changes—for example, mammals as opposed to dinosaurs beyond the Cretaceous-Tertiary boundary.

Ultimately, the tree of life would seem to grow by Lenin's rule of revolutions: one step forward for two steps back. Death may clean up the detritus of history, but it does not advance history. One does not die at the behest of population dynamics, and death is not adapted to making room at the top.[13]

Do Species Have a Finite Life Span?

For us, a life span—the interval between fertilization and death—is frequently confused with a lifetime—the interval between birth and death. Be that as it may, the question here is whether an average or even a maximum lifetime is determined in our species. In other words, is life span or lifetime a species-specific characteristic or merely a circumstantial characteristic, possibly species-typical but without any causal connotation of built-in limit?

Life spans are described in several ways (mean, median, mode, etc.) and are visualized in different ways. For present purposes, the most convenient way of illustrating life spans is as survivorship distributions, the rate at which a cohort (all the organisms starting their life span at the same time) dies out. Survivorship curves demonstrate the totally different ways cohorts of different species die out while living in their different environments and making their living in different ways.

The conjunction of the surviving number of organisms in a cohort (along the Y axis) and the period of time (along the X axis) until the last member of the cohort is dead is plotted in a survivorship distribution. In the distributions shown in figure 1.1, the time axis is calibrated in fractions of a life span (centiles or hundredths of a lifetime) in order to facilitate comparisons between species with differing life spans.

The three species with survivorship distributions plotted here are *Homo sapiens,* represented by a 1910 cohort of white males (open squares), a ubiquitous, microscopic rotifer, *Proales decipiens* (closed triangles), and the fruit fly *Drosophila* (closed diamonds) captured in the wild.[14] The distributions illustrate how death erodes each cohort under natural conditions (as opposed to the artificial and virtually sterile conditions of the laboratory).[15]

Each of the three distributions has an inverted S shape, beginning and ending with more nearly flat portions connected by a smoothly curving diagonal portion. The flattened portions at the beginning indicate how long members of a cohort live before death begins to take its toll, while the flattened portions at the end indicate how long members of a cohort live before death completes its job in old age. The curvature in the middle portions is a function of how rapidly death descends upon the cohort between an initial delay and a late deceleration.

Were species-determination to play no part in influencing individuals' life span and were death entirely a random event, the survivorship distribution would fall off at a constant rate throughout the distribution. Alternatively, were species-determination the sole influence on individuals' life spans, the survivorship distribution would be maximal and level at first, and then would drop precipitously down to zero at the age when individuals reach their species-specific life span. Of course, in a state of nature, or even in a laboratory, organisms may die from vague causes that distort a distribution, including statistical error in collecting data and random deviation from the ideal.. These nebulous causes must be accepted without clouding a view of the principal causes of death.

The curve for wild *Drosophila* comes nearest the prediction for death as a function of random accident with constant probability, but even this curve

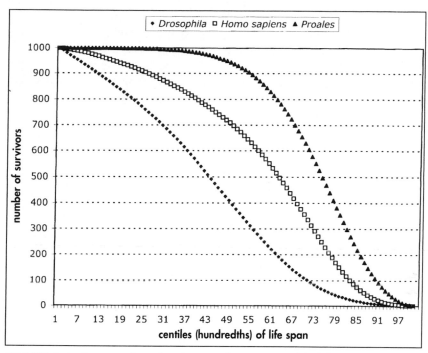

FIGURE 1.1. Survivorship distributions for comparable life spans. (Curves drawn from data in Pearl, 1924, 376–77, table 112.)

bends slightly at the beginning and levels off conspicuously as the cohort's membership approaches zero. *Drosophila* raised under laboratory conditions produce survivorship distributions virtually identical to those shown here for human beings, suggesting that animals in nature suffer from a number of diseases that are not present under conditions of domestication.

In contrast to the *Drosophila* distribution, the distribution for the survival of the tiny rotifer, *Proales,* comes nearest the prediction for death as a function of a species' life-limit, proceeding nearly horizontally at first before dropping off dramatically. The survivorship distribution for *Homo sapiens* is intermediate: somewhat flat at the beginning before dipping and flattening at the end as the death rate slows.

Thus, while random accidents may play a nearly constant role in killing off *Drosophila* in the wild during most of their lifetime, accidents play a minor role in killing off *Proales* and an intermediate role in killing off *Homo sapiens.* On the other hand, the life span of *Proales* would seem very much more biologically determined than the life spans of *Drosophila* and *Homo sapiens.* The tiny rotifer would seem, somehow, to die on a schedule, with the absolute duration of its life span (that is, its life-limit) strongly determined. The duration of a life span in *Homo sapiens* would seem less biologically determined than *Proales* and that of *Drosophila* would seem least determined of the three.

Of course, biological determination is influenced by many things, from genetics to epigenetics, from nuclear genes to environmental effects, and one must always bear alternatives in mind, as well as their possible interactions, when speculating on biological determination. But, mutations altering life span–determination in the three species would be expected to alter the survivorship distributions differently to the degree that genes alone influence biological determination. Thus, in the case of *Proales,* mutations affecting the life span might delay the onset of death, thereby extending the interval of life. In *Drosophila,* mutations affecting the lifetime might create more resistance to disease, pushing the survivorship curve upward (rounding the straight line).[16] In *Homo sapiens,* different mutations affecting the lifetime might change both parameters: push the survivorship curve upward and extend its limit.

Actually, selection for eggs, but not mutants as such, of young rotifers extends life span,[17] and gerontologist Caleb Finch suggests that rotifers "give a model for the relationship between specific cytoplasmic determinants during oogenesis and the epigenetic control of senescence."[18] On the other hand, mutations, rather than epigenetic controls, would seem to be involved in the lengthening of lifetime in the roundworm, *Caenorhabditis elegans*, when too much of the protein Sir2 (silent information regulator 2) is produced in mutants.[19] The evidence in *Drosophila* and mice is, however, ambiguous, since the lengthening of lifetime in fruit flies may be spontaneously reversed, possibly by affecting development, and, in mammals, genes affecting life span also influence growth and cause cardiopulmonary lesions as much as influence aging.[20] The effects of mutants on the average human life span are simply uncertain, and gerontologists Leonid Gavrilov and Natalia Gavrilova warn that "the age-dependent component of mortality . . . is historically stable."[21]

The species-specificity of biological "destiny," thus, would seem to work differently in *Homo, Proales*, and *Drosophila*. These organisms evolved under different circumstances and with different histories, producing different overall strategies for life, for survival, reproduction, and death. If one ignores for the moment all the complexity that goes into evolution, notably fecundity, the rotifer would seem narrowly determined to get it over with, while the fruit fly takes its chances, and the human being hedges its bets.

In effect, genetic, epigenetic, and environmental effects all come to bear on biological determination and one cannot exclude any of these influences. One could do little to effect change in the rotifer's lifetime without changing its species-specific biological determinants (whether genetic or epigenetic); the fruit fly's lifetime could be changed most rapidly by changing its environmental exposure or its intrinsic fragility (that is, eliminating the kinds of events that kill it or render it vulnerable to these events); the human being would fall somewhere between, subject to both rapid change due to local circumstance and long-range change due to changing its biological nature genetically or epigenetically. In any event, unlike *Proales,* our species-specific determinants are not our main executioner. The life span of *Homo sapiens* may, indeed, be species-typical (or what are statistics for?), but neither an average nor a maximum would seem species-determined.

DO WE DIE AT THE COMMAND OF KILLER GENES?

Ever since 1953 when James Watson and Francis Crick succeeded in reducing genetic continuity in deoxyribonucleic acid—better known as DNA—to the simple game of matching base pairs (adenine [A] to thymine [T] and cytosine [C] to guanine [G] or A → T; C → G), genetics has dominated the life sciences. Indeed, reducing biological complexity to its genetic components is the predominant objective, if not the only objective, of most research in the life sciences and the *raison d'être* of the multinational, multibillion–dollar Human Genome Project.

But genes are not the only things that influence heredity. We are constantly learning about other influences, from mitochondria to DNA methylation, all of which fall vaguely and loosely under the umbrella of *epigenetic controls,* reprogramming or specific changes to the *epigenome.* Indeed, in addition to "the major type of DNA modification . . . [via] the methylation at cytosines, there are multiple modifications associated with chromatin . . . [in which] hereditability has been demonstrated only in rare cases."[22] Gerontology is, however, so deeply imbued with biology's genetic paradigm that virtually any other approach to solving the problems of aging, dying, and death is rejected and tarred with the brush of holism (antireductionism) if not vitalism.

What is it, then, that genes could do to influence our life span, our aging, our dying, and our death? In general, genes work through their products, frequently ribonucleic acid (RNA) and hence proteins. Even the most far-reaching genes, those that determine hereditary traits, have their most immediate effects within the cells that produce the gene's coded RNA and resulting protein. In turn, the products of cells operate on tissues, organs, and organ systems by interactions, through induction and transduction pathways. The products of genes may operate at one stage of development or throughout the course of a lifetime, in everyday upkeep, and/or in response to challenges. But in every case, genes are thought to exert their influence through some effect on cells or their products, and cells then mediate the indirect effects of genes.

How, then, could genes intervene in life spans and cause aging, dying, and death? Ordinarily, cells in many tissues throughout the body undergo turnover: differentiated cells die and are replaced by new cells. At one time, one would have said that the cells die in the course of differentiation, for example, in the case of the keratinizing epidermis, but today, cells are said to die through programmed cell death (PCD) involving one or another mechanism: apoptosis, in which single cells die and are digested by so-called macrophages; and autophagia, in which groups of cells dissolve or harden (i.e., tan) under the influence of their own lytic enzymes or denaturing mechanisms. Specifically, genes said to be involved in aging are widely thought to operate through cumulative effects on cell loss over time, especially cell loss implicated in disease (for example, neurodegeneration, retinal degeneration, cardiovascular disease) and increased frailty or vulnerability to a variety of diseases.[23] On the other hand, genes said to be involved in life's prolongation are thought to operate by attenuating the loss of cells. Thus, for example, "long-lived genetic mutants such as the p66[sch] knockout mouse are typically less prone to stress-induced apoptosis [than normal mice]."[24]

Aging, Dying, and Death Genes

The possibility of genes governing aging, dying, and death has a number of permutations. There would seem to be no end of genes that influence life span.[25] The gerontologist Tom Kirkwood has proposed, under the title of the "disposable soma" hypothesis, that organisms, especially long-lived, complex organisms, employ considerable numbers of genes in regulative roles supporting growth, development, and maintenance. Aging results from the accumulation of irreparable defects in these genes and hence in the failure of cells to maintain and repair the soma (body) in the wake of stress and environmental hazards.[26]

The authors of *Successful Aging,* John W. Rowe and Robert L. Kahn, are slightly more circumspect:

> [T]he strongest influence of heredity on aging relates to genetic diseases that can shorten life, such as numerous forms of cancer and familial high cholesterol syndromes (which lead to heart disease). . . . Still, however, heredity is not as powerful a player as many assume. For all but the most strongly determined genetic diseases, such as Huntington's disease, MacArthur Studies show that the environment and lifestyle have a powerful impact on the likelihood of actually developing the disorder. . . . Genes play a key role in promoting disease, but they are certainly less than half the story.[27]

The bio-gerontologist Aubrey de Grey goes further: "Genes are not responsible for aging. Genes are responsible for defending us, to a greater or lesser degree depending on the species, AGAINST aging."[28] Moreover, according to the gerontologists Jay Olshansky and Bruce Carnes, "[t]he requirement that death genes become activated at ages beyond the reproductive years means that evolution could not give rise to them."[29] And the science writer Stephen Hall quotes the gerontologist Leonard Hayflick, the grandparent of all cell-aging studies, as insisting that "[t]here are no genes for aging . . . I'll say that categorically, and I'll defend it despite what you have heard."[30] Natalia Gavrilova and Leonid Gavrilov state equally categorically that "many of these 'self-evident' assumptions (for example, the normal life span distribution law, and the notion of an absolute limit to longevity) are simply unsound when tested . . . and an absolute upper limit to longevity appears not to exist."[31]

The obvious problem with genes for aging, dying, and death is that they would seem to offer no adaptive advantage to individuals possessing these genes, and, hence, would have no way of evolving into stable parts of the genome. Modern genetics may attempt to rescue death genes as hitchhikers or deceivers, but the attempts are unconvincing. Deleterious genes may get into the genome by hitchhiking—going along for the ride, so to speak—were they closely linked to adaptive genes, but no such hitchhikers are presently known. Moreover, genes getting into the genome by deception might enhance the fitness of the individual at one stage of life only to diminish fitness at another stage, but why would the same gene have opposite effects at different times of life?

The evolutionary biologist George Williams's "theory of antagonistic pleiotropy" is a theory of genetic deception. "Pleiotropy" refers to genes with more than one effect, while "antagonistic" implies that these effects are contradictory. The theory would have the pleiotropic effects occurring serially, and thus the effects follow one another. Williams suggests that a net gain in Darwinian fitness would accrue to organisms were genes with favorable effects prior to or during the reproductive period of a lifetime to have deleterious effects in the late or postreproductive period.[32] Attributing opposite effects to genes for the sake of explaining aging would seem circular, but many gerontologists find the theory of antagonistic serial pleiotropy attractive and continue looking for once felicitous genes that become deleterious and cause aging, dying, and death late in life. Certainly, all of biology will take notice if these gerontologists come up with some such genes, but, at present, the search has been fruitless.

The Sad History of Longevity Genetics

Genetics' importance for biology begins long before Watson and Crick with the "rediscovery" of Mendel's laws of hereditary at the beginning of the twentieth century. Since then, biologists have been divided between those who attempt to analyze life as something determined by genes and those who concede that the mixture of genetic and environmental factors are inseparable.[33] (Those who suggest that non-Mendelian heredity may also play a role may be making a comeback,[34] but those who might have argued in favor of purely environmental determinants of life have long since been drummed out of the profession.) Of course, a great deal of the debate between members of the two camps hinges on exactly what one means by genes, but the definition of genes has only become more confused and controversial with the passage of time.

For twentieth century evolutionists, the foremost problem that Mendelian genetics was supposed to solve was how Darwinian evolution by natural selection worked at the level of genes.[35] But, for the first quarter of the twentieth century, Mendelian genetics failed to illuminate evolution at all. Many of Darwin's most loyal supporters took different and competing sides of the issue. The embryologist-turned geneticist Thomas Hunt Morgan and his coterie in the "fly room" laboratory at Columbia University became the strongest adherents to the strict Mendelian precept of particulate inheritance. Morgan examined qualitative inheritance and largely ignored natural selection's requirement for the inheritance of small, quantitative changes. The Dutch botanist Hugo DeVries showed how a rare, large, hereditary change, called a *mutation*, could create virtually new species in a single step, but his discovery was so antithetical to the gradualism of natural selection that it threatened to scuttle Darwinism altogether. The equilibrium discovered by Goddfrey Harold Hardy and Wilhelm Weinberg, and known as the Hardy/ Weinberg law, moreover, demonstrated that infrequent mutations could have only minimal effects on populations. Meanwhile, William Bateson, the Cambridge zoologist and "apostle of Mendel"[36] who coined "genetics" but not the "gene,"[37] floated a version of Mendelian factors at odds with both Morgan's chromosomal theory and the notion of quantitative inheritance spawned by the London biometrician, Karl Pearson.

Among the early geneticists, Pearson was most interested in longevity and might have kick-started the study of longevity's inheritance had his reputation not been sullied by his penchant for eugenics and had he not been denounced as anti-Mendelian by Bateson. What Pearson established and legitimized was the way to study biometric traits, such as height, weight and longevity, through distributions, and he effectively invented population statistics in order to study distributions (although Francis Galton is usually given the credit). When the frequency of a biometric trait was found to have a normal, bell-shaped distribution, Pearson argued, some biological constraint determined the mean (the vertical line at the center of the bell), while small variations expressed among members of a population and the chance of the draw explained the error or scatter of points around the mean (the area beneath the bell on either side of the mean). The mean and scatter, in terms of the standard deviation of the mean, provided a basis for describing and comparing distributions, but in the early days, attempts to define the "significance" of differences was left to "good judgment."[38]

Pearson proceeded to work out a mathematics of skewness—the asymmetry of a distribution favoring one side or the other—when things got lopsided and the mean (average) and mode (most common value) did not match. Pearson proposed dissecting skewness by identifying normal curves within observed distributions. Pearson should also be credited with introducing biologists to the study of distributions, inventing variance and the standard deviation to describe scatter, and devising the chi square method for evaluating statistical differences.

Regrettably, Pearson's biometrics hardly got off the ground, and he did not establish curve analysis as a standard instrument for studying longevity. Instead, quantitative genetics replaced biometric analysis when Ronald Fisher, J. B. S. Haldane, and Sewall Wright packaged genetics and natural selection together with literary and mathematical eloquence in a new synthesis, followed by Theodosius Gregorievitch Dobzhansky's "New World" synthesis or "synthetic theory" of evolution, and Julian Huxley's "modern synthesis," launching the reign of still-fashionable neo-Darwinism. Darwinian evolution was thus rescued from the junk heap of unproven hypotheses, but at the same time, the study of heredity was directed toward (reduced to) the Morgan style of particulate genes on chromosomes and away from the Pearson style of curve analysis. The difficulty geneticists had explaining why biometric distributions were smooth rather than stepwise to meet the requirements of qualitatively discrete genes was soon rationalized as the environments' ability to burnish rough edges and as statistical error surrounding additive effects of quantitative genes.

Model Systems

Genetics has proved an overwhelming boon to the fortunes of biology. Virtually any research project stated in genetic terms will be funded by a governmental or nongovernmental agency. Thus the genetics of aging, dying, and death are widely studied in so-called model systems, namely, budding yeast, *Saccharomyces cerevisiae (S. cerevisiae)*, the roundworm, *Caenorhabditis elegans (C. elegans)*, the fruit fly, *Drosophila melanogaster (Drosophila)*[39], and, since the advent of patented, bioengineered mice, in the laboratory mouse, *Mus musculus*.

The overwhelming advantage of working with model systems has been apparent since bio-gerontologist Raymond Pearl's classic work on fruit flies,[40] namely, model systems allow the experimenter to use laboratory reared, genetically homogeneous organisms (and throw away the organisms without pangs of conscience after performing experiments). In addition, the organisms chosen for model systems are highly fecund and have short generational times, making the study of aging that much easier and cost efficient compared to waiting around while a slowly reproducing and slowly aging organism responds to experimental manipulation. But the experimental genetics' approach to longevity research in model systems would not have gotten to first base if it had not shown that "remarkable life-span extensions can be produced with no apparent loss of health or vitality by perturbing a small number of genes and tissues."[41] Although this quotation is borrowed from a study on the roundworm, similar conclusions are drawn from work on yeast, flies, and mice.[42] Indeed, these model systems are said to have turned up a number of "mammalian gerontic genes (those specifically associated with the aging process)."[43]

No doubt, genes can influence life expectancy or aging phenotype. Some genes or mutations expand life expectancy, if at a price by way of competitive disadvantage,[44] and some genes shorten life expectancy through a variety of mechanisms.[45] Caleb Finch testifies in favor of "inarguably, programmed senescence," citing, as his exemplar, genes determining "deficient mouthparts . . . [of insects with an] adult phase of 1 year or less."[46] For example, the ultra-short life of some adult mayflies (literally minutes to a few weeks) is correlated with the insect's genetically determined aphagous anatomy.

And mutations determining abnormal anatomies may also affect longevity. For example, in *Drosophila*, a mutant gene known as *vestigial,* which causes shriveling of wings, also causes premature death. The average life expectancy of female and male flies expressing *vestigial* is reduced 41 and 31 percent, respectively. But whether *vestigial* is a gerontic gene is another matter. Rather, *vestigial* would seem somehow to have affected anatomy and, only secondarily, the aging process.

On the theoretical side, the chief problem faced by gerontologists trying to assess the role of longevity genes in model systems is identifying genes affecting universal aging processes rather than species-typical processes. For example, as pointed out in a recent review of progeroid syndromes in human beings, "in *D. melanogaster* females, . . . a major cause of aging and death is the toxic effect of compounds present in the seminal fluid products secreted from the male fruit fly accessory gland. . . . [These compounds are] not considered a primary cause of mammalian aging. Similarly, . . . replicative senescence (the loss of divisional capacity in the mitotic tissue compartments of the soma) is not a potential aging mechanism for organisms whose soma are completely postmitotic, such as *C. elegans*."[47] Later, the authors point out that *C. elegans* dies of extreme cuticle thickness and *S. cerevisiae* of extrachromosomal ribosomal DNA circles, neither of which mechanism would seem of universal relevance or particular importance to human beings.[48]

On the practical side, the chief problem posed by genes in model systems would seem to be specificity: that *Homo sapiens* is *Homo sapiens* and not *S. cerevisiae, C. elegans, D. melanogaster,* or *Mus musculus*. As demonstrated above, the survivorship curve for *Homo sapiens* has its own species-typical shape, suggesting that *Homo sapiens* is adapted to its own, species-typical niche, which, if not unique, is undoubtedly different from the niches of the chief model-systems. Even de Grey, who asserts that "[i]t is to be expected that aging of rather distantly related organisms will share fundamental characteristics," also acknowledges that the same organisms "will fail to share more secondary characteristics—just as is in fact seen."[49]

One is not surprised that the survivorship distribution for *Drosophila* (and one might add *S. cerevisiae*) can be blown upward from virtually straight diagonal lines to complex inverted S shapes through the manipulation of environments, and one cannot doubt that selective breeding can result in both lengthening and shortening longevity in *C. elegans* by enhancing or inhibiting lethal and deleterious effects of genes. Clearly, the short-lived model systems currently under study are appropriate for their intended purpose—aiding the study of qualitative, longevity genes—and they have been eminently successful for discovering such genes. It is only the relevance of these genes to human aging, dying, and death that is questionable!

Human Studies of Longevity's Genetic Controls

Several direct approaches have been taken to determine genetic contributions to longevity in human beings. The traditional approach evaluates pedigrees and familial correlations at the age of death. For example, one would be tempted to conclude that inheritance played a large role in the case of the extraordinary longevity of Jeanne Calment, who died at 122+, since her "direct forbearers . . . lived on average 80 years compared to only 58 years for the ascendants of other members of her family of the same generation."[50] The problem with pedigree studies is translating them from mere anecdotes without quantitative prospects into serious efforts to identify genes with definitive roles in longevity. Efforts to solve this problem are traced by Raymond Pearl in *The Biology of Death* and, with his daughter, Ruth DeWitt Pearl, in *The Ancestry of the Long-Lived*.

The idea of pedigree and familial correlations is deceptively simple: if heredity plays a part in longevity, those with the greatest longevity should be the offspring of long-lived or "longevous" parents and the parents of longevous progeny. Karl Pearson and Miss Beeton (*sic*) performed the first test of this hypothesis using the technique now known as meta-analysis. Together they gleaned data from published records of the peerage, the landed gentry. These data covered the ages of fathers and sons at death and brothers dying beyond the age of twenty. Later, records from the English Society of Friends and the Friends' Provident Association were added to the analysis in order to study deaths of female relatives and infants. All these data on age at death were paired for parents and offspring (direct lineal inheritance) and for offspring of the same parents (collateral inheritance); the coefficient of correlation—the degree of mutual dependence—was calculated for each pair, and statistical significance was assessed by comparisons to the probable error. All the correlations judged to be significant were positive, meaning that the life spans of parents and offspring increased in unison.

Alexander Graham Bell then studied the Hyde family in a similar way. Of 767 offspring who lived to eighty years or more, 48 percent had parents who lived to eighty years or more. In Pearl's words, "there is a definite and close connection between the average longevity of parents and that of their children." Pearl, then, strikes a proverbial note in summarizing Bell's finding: "[A] careful selection of one's parents in respect of longevity is the most reliable form of personal life insurance."[51]

According to Bell's data, longevous parents add as much as twenty years to the average life span of their offspring. These twenty years would correspond to the contribution of genes to longevity. Similarly, if not quite, according to a canvass of prominent physicians at the time, longevous parents would add about thirteen years to the average life span of offspring if diseases encountered in a lifetime are factored out (based on the mortality experience of 1900–1910).

But all is not well with correlation coefficients in the study of the heritability of longevity. Indeed, rather than extreme long life running in families, "[t]he extremely longevous person tends to be exceptional, even in his own sibship."[52] Working on an extensive data set of parent-offspring correlations, Pearl and DeWitt Pearl concluded, "that the biometric method of correlation, as it has hitherto been applied to the problem of the inheritance of longevity, is an inadequate and unreliable method."[53]

The overriding problem with pedigree and familial correlation studies is that genes for normal longevity (as opposed to genes for progeria, Hutchinson-Gilford syndrome, Werner syndrome, and other congenital disorders) have never been successfully associated with either discontinuous variables, that is, qualitative (Mendelian) genes, or with continuous variables or quantitative (poly-)genes. In effect, pedigrees may not be tracing genes as such. But all is not lost: this problem is confronted (if not overcome) by twin studies, which make it possible to draw distinctions between environmental and genetic effects on heredity.

Unlike pedigree and familial studies, twin studies offer a direct approach for estimating the dimension of genes's role in longevity. The relevant variable is called "life span heritability," the proportion of variance among individuals, at the age of death that is attributable to differences in genotype. Life span heritability is ascertained in twin studies by comparing the mortality rates and age at death for twin-pairs, both identical and like-gender fraternal twins, including twins reared apart, as well as brothers and sisters in the remainder of the population. Surprisingly, a Danish twin study concluded that "longevity seems to be only moderately heritable," with a genetic component no greater than 26 percent for males and 23 percent for females.[54]A Swedish twin study found that any genetic effect was small, or even absent for males.[55] With percentages such as these—closer to 0 than 100 percent—notions of genetic control over maximum life span in human beings are hardly robust and persuasive. Indeed, the demographer Väinö Kannisto concludes, "The heritability of longevity . . . is very weak."[56]

Is Longevity Ultimately Inherited?

Whether one considers longevity inherited or not will depend on what is meant by "inherited." One will have a different answer if one interprets "inherited" to mean strictly by Mendelian genes as opposed to all the other influences—epigenetic and environmental—that impinge on heredity.

Longevity is certainly genetic, but in the special sense that genes operate against alternatives. Genes set many biometric parameters in this negative way. For example, genes determine that we are not, on average, eight feet tall and do not weigh five hundred pounds. Our species' genes resist these possibilities. But this is not to say that we possess genes that determine our average height or weight. Likewise, at present, in developed countries, half of us will live to about 80 years and not to 120 years. This is not to say that we have genes for an 80-year lifetime, but genes would seem to militate against our living to 120 years.

Beyond Mendelian genes, many biological attributes bear some relationship to the inheritance of longevity. For example, small mammals with high metabolic rates, such as mice and rats, live relatively short lives compared to large mammals with relatively low metabolic rates, such as horses, humans, and bowhead whales. But none of these parameters determine longevity any more than genes. Indeed, *"[t]here is no generally valid, orderly relationship between the average duration of life of the individuals composing a species and any other broad fact now known in their life history, or their structure, or their physiology."*[57] Even metabolic rate gives no reliable clue to life span generally. Bats, for example, have higher metabolic rates than mice and rats but live relatively long lives. Similarly, birds with high metabolic rates live longer than mammals of comparable size and low metabolic rates. Likewise, other biometric parameters—body size, weight, brain size, brain size–body weight ratios—would seem to have a bearing on longevity in some species but not in others. Indeed, no amount of shuffling data has demonstrated an unambiguous correlation of longevity with any biological attribute.

Possibly, bio-gerontologists are looking for the wrong sort of thing in their quest to attribute longevity to heredity. Could longevity exhibit non-Mendelian inheritance? The correlation of offsprings' longevity with the male parent's age points in that direction (see chapter 5 for further details). According to a *Sidney Morning Herald* journalist who covered a recent international longevity conference, "Research from the University of Chicago's Centre on Ageing shows that daughters born to fathers in their late 40s or older live, on average, three years less than other women, yet their brothers are not affected. . . . But the answer is not to leap into fatherhood early in life, because daughters born to fathers aged under 25 also have a shortened life span, said the center's research associate, Natalia Gavrilova."[58]

Natalia Gavrilova also looked at links between long life and motherhood: "We found that, in contrast to previous reports by other authors, women's exceptional longevity is not associated with infertility. . . . There is no relationship between childlessness and longevity."[59] Gavrilova may have taken her cue from the science-fiction writer Bob Shaw, who portrays a society of impotent immortal males but perfectly fecund immortal females.[60] Other, more fruitful, avenues for research may lie ahead, but let us lay to rest the notion of killer genes. In sum, we do not die, at least not directly, at the behest of any gene.

Do We Die at the Command of Killer Environments?

There is, of course, no end of things in our environment that can kill us, from accident to pollution and from trauma to infections. That's not the problem. The problem is that, like genes, we cannot live without our environment. Life is a compromise with both our genes and our environment. There is no such thing as perfection. We simply make do with what is at hand, although we might wonder if other environments, like other genes, might keep us alive longer and better.

Environments enter mortality statistics in two ways: causes of premature death and promoters of aging. Regrettably, experimental gerontologists working at the genetic/molecular level are prone to confuse these environmental influences, for example, when arguing that "[o]ur ability to rapidly stockpile energy during periods of abundance and to conserve energy during times of famine . . . [are] ill suited to the sedentary lifestyles and rich diets of modern society."[61] No doubt, we are exposed to lots of hazards through our interactions with our environments, and becoming a "couch potato" is dangerous to our health and should be resisted or avoided, but causes of premature death, like the proverbial Mack truck, do not necessarily promote aging. Soldiers are killed by hostile and friendly fire while waging modern war, but civilians, especially children, are killed by disease, malnutrition, and neglect, none of which qualify as promoters of aging.

So much of aging involves our interactions with our environment that the environment is inevitably one of the usual suspects determining aging, dying, and death. For example, we blame close work and the sun for presbyopia (loss of close vision), loss of accommodation, cataracts, and macular degeneration, and, more seriously, mutagenic effects.[62] We also blame loud music and jackhammers for presbycusis (loss of hearing in the high-frequency range). Moreover, strains of work lower our general level of motor activity and decrease our fine motor skills, and boredom destroys our capacity for running memory.

The strongest cases for a direct environmental influence on longevity are made by the near universality with which lowering temperature (hypothermia) in poikilotherms (cold-blooded animals) including fish, and imposing a regime of caloric restriction (CR), also known as nutritional restriction (NR) and dietary restriction (DR), in a host of organisms including homeotherms (warm-blooded animals), prolongs longevity.[63] The effects of hypothermia on longevity seem to be mediated by influences on "maturation, [and] adult metabolism,"[64] while the effects of caloric restriction are thought to lower "mortality entirely as a consequence of a lower short-term risk of death."[65] These environmental effects may work through any or all of several mechanisms: by having "a protective effect . . . on fuel use"[66] through lowered plasma levels of both glucose and insulin; by inducing hyperadrenocorticism with "an effect over the lifetime similar to that of the transient acute hyperadrenocortical response to stress . . . [serving] as a buffer, [and] keeping primary defenses such as inflammatory and immune (including autoimmune) responses in check."[67] Caloric restriction, thus, could postpone or prevent "a remarkable array of diseases and age-dependent deterioration, without causing irreversible developmental or reproductive defects."[68]

The possibility that hypothermia and caloric restriction work through the same mechanism or parallel effects on metabolism is difficult to test, since homeotherms are not good subjects for hypothermia experiments. But species of mammals exhibiting natural torpor or hibernation sustain decreased body temperatures and "live unusually long in relation to their specific metabolic rate when active,"[69] suggesting that hypothermia and caloric restriction meet on the same epigenetic pathway. In particular, hypothermia and caloric restriction would seem to be linked via stress.[70] Chronic stress is typically thought of as accelerating the onset of senescence or aging, but stress also implies pressures and tensions on metabolic regulation, reproductive control, the inhibition of cellular proliferation, and the promotion of programmed cell death—all of which are relevant to the underlying bio-molecular pathways of longevity (DNA repair, oxidative stress response, release of microbicidals, and so on). In the C. elegans, the molecular responses triggered by environmental stress are even called "the transcriptional equivalent of the fountain of youth."[71] In other words, environments and biological determinants cannot be separated. We no more die at the command of our environment than we die at the command of our genes.

Is Irrepressible Entropy Our Executioner?

Do we die at thermodynamics' command? Do the laws of thermodynamics that rule the universe also rule our life span? Biologists, gerontologists, and physicians, with a reductionist physical/chemical bent, have brought life and death under the umbrella of the laws of thermodynamics, contending that life is inevitably under threat because nothing dissipating energy can escape degradation! But is this scientific argument for the certainty of death compelling? Is belief in thermodynamics any more persuasive than believe in the "immutable" laws of theologians in transcendental power or the power of species, genes, and environments to kill?

Many dedicated scientists will say that death is inescapable, because we live in a thermodynamic universe in which everything rolls down an energetic hill. According to the laws of thermodynamics, in a thermodynamic universe, nothing mechanical—including living things—can operate and remain unchanged in perpetuity or ever return to an originally pristine condition. In other words, everything that uses energy ultimately runs down, and, when living things run down completely, they return to dust—nonliving stuff.

Other gerontologists have a problem with this point of view. Indeed, Raymond Pearl concluded: "A death really due to . . . a breaking down or wearing out of all the organ systems of the body contemporaneously . . . probably never, or at least extremely rarely, happens."[72] Who is right?

Thermodynamics, the branch of mechanics concerned originally with heat's movement in steam engines—hence the "thermo-" and "-dynamics" in thermodynamics—provided the theory that mastered steam and powered the industrial revolution. But thermodynamics did not stop there. Standing on the shoulders of seventeenth century giants Boyle and Newton, with temperature, pressure, and volume to guide their study of mechanical action and work, the great eighteenth and nineteenth century physicists and engineers—Boltzmann, Carnot, Clausius, Evans, Gibbs, Joule, Kelvin, Rankine, Trevithick, and Watt—devised the laws that moved beyond steam to other forms of energy and beyond boilers and pistons to other forms of engines and machines. Ultimately, the laws of thermodynamics were perceived to rule the universe: the total amount of energy in the universe is constant, but the inaccessible (useless) part of this energy tends to increase.

The first law of thermodynamics, the conservation of energy, states that energy is neither created nor destroyed but is only transformed, for example, while performing work. Strictly speaking, the law applies only to closed systems, which is to say, systems encompassing all the relevant energy whose transformation is taking place, and, thus, systems not exchanging energy with other systems. The law, therefore, would seem to leave out living things, since, if nothing else, living things are forever exchanging energy with their environments. For example, every time you breathe you are taking a source of potential energy (in the form of oxygen) from your environment and returning a product of spent energy (in the form of gaseous carbon dioxide and water) to your environment. But the laws of thermodynamics are generally considered so universal in their applicability that caveats regarding closed systems are ignored.

Originally, the second law was concerned with heat: heat does not run upward by itself, that is, pass from a body at lower temperature to another body at a higher temperature. Later, when a correlation was suspected between heat and the movement of particles in bodies, the law was restated in terms of populations of moving particles and statistical probabilities: slower particles will not on average move to bodies of faster moving particles.[73] Later still, the law was restated even more broadly in terms of organization: disordered associations of particles do not spontaneously pass order to more ordered associations of particles. On the contrary, whenever work was performed and free energy was transformed, a part of it was dissipated and rendered disorganized. This part was called *entropy* (coined by Clausius in 1865). Thereafter, free energy or the potential for performing useful work was equated with order, and entropy or the unlikelihood of performing useful work was equated with disorder.

Thus, according to the second law, ordered associations become disordered in the process of performing work. As work is performed, free energy is dissipated and entropy increases until the system reaches the point known as thermodynamic equilibrium, where free energy is at its minimum and entropy is at its maximum. As engines, or systems that do work, approach thermodynamic equilibrium, they do not merely run out of steam; they break down, and when they reach thermodynamic equilibrium, they break down completely.

The situation regarding living things would, therefore, seem starkly clear: as an ordered structure performing work (that is, metabolism, synthesis, etc.), a living thing becomes increasingly unstructured and in/operative in the course of a lifetime, until, at death, it turns into a corpse with less structure and activity. Entropy may then reach its maximum, and at thermodynamic equilibrium, the once-living thing is randomly distributed dust.[74]

This second law is considered so utterly incontrovertible that phenomena, including life, are routinely assumed to operate under its aegis. Indeed, chemists, physicists, and biologists are so completely convinced of the universality of the second law that there would hardly be modern science without it. The Nobel Prize–winner and widely credited co-discoverer of messenger

RNA (mRNA), François Jacob, was not overstating when he wrote, "It is no exaggeration to say that the way we now regard nature has to a large extend been fashioned by . . . thermodynamics, which has transformed both the objectives and the outlook of biology."[75]

Problems with Thermodynamics

Thermodynamics is like a great tranquil pool covering an immense scientific terrain, but a few irregularities remain under the smooth surface. Philosophers have had problems with what would seem to be an inherent degree of circularity or illogic, and a tautology or a redundancy in thermodynamics' law. Are the energies extant in our universe sufficiently represented in thermodynamics or has "the measurement of each [kind of energy] . . . been so chosen as to justify the principle of conservation of energy"?[76] Is the second law truly applicable to life or is it a "hypothesis . . . as irrefutable as it is indemonstrable"[77]?

At issue is reversibility, banned by the second law. Certainly, irreversibility is a commonplace, and we have all experienced something or other breaking down beyond the point of repair. On the other hand, reversibility is also a commonplace, especially in the quotidian world of life. We see reversible events happening every day: when we awaken in the morning, in the springtime, when a baby is born, when we shave or cut our hair, and so on. The question needs to be asked: does reversibility make life different from everything else and place life outside the realm of the second law? Are the laws of thermodynamics an explanation for the breakdown of living things or merely for the breakdown of dead things? Do the laws explain the reversibility of life and the irreversibility of death?

Philosophers have had problems with thermodynamics' failure to take into account life's and evolution's creative powers. The philosopher of becoming, Henri Bergson, in particular, confronted thermodynamics' block to creativity and suffered the price for his effrontery by being charged by scientists with vitalism and dispatched beyond the pale of scientific notice. For Bergson, "the direction, which this [thermodynamic] reality takes, suggests to us the idea of a thing *unmaking itself* . . . [In contrast] life [is] an effort to re-mount the incline that matter descends. . . . The life that evolves on the surface of our planet is indeed attached to matter. . . . In fact, it is riveted to an organism that subjects it to the general laws of inert matter. But everything happens as if it were doing its utmost to set itself free from these laws."[78] Bergson concludes, "In vital activity we see . . . *a reality which is making itself in a reality which is unmaking itself*" (emphasis in original).[79] The challenge for bio-gerontologists is to flesh out that "*reality which is making itself*," not to ignore it.

The philosopher Martin Heidegger seems to have reduced this problem to a mere predicament by distinguishing between technology and the essence of technology and between the instrumental and anthropological definitions of technology. Heidegger begins by arguing that the "instrumental definition of technology is correct . . . [However,] the merely correct is not yet the true." And, he goes on, "Technology is a way of revealing . . . truth," and of "[u]nlocking transforming, storing, distributing and switching about . . . [but revealing truth] never simply comes to an end. Neither does it run off into the indeterminate."[80] Heidegger acknowledges the human will to exploit the energies of nature, but he couples that will to a willingness to follow the challenge of revealing, even if modern physics (including thermodynamics) "must resign itself ever increasingly to the fact that its realm of representation remains inscrutable and incapable of being visualized."[81]

The problem of visually representing thermodynamic equilibrium as randomness was brought up again, more recently, by the applied mathematician Stephen Wolfram. He wondered if randomness were quite as disordered as is generally supposed. Indeed, there may not even be such a thing as randomness at all, but, if there is, it is an order rather than a disorder. According to Wolfram, "when we say that something seems random [i.e., has maximum entropy] what we usually mean is that there are not significant regularities [for us to] discern." He continues: "[I]t is easy to generate behavior in which our standard methods of perception and analysis recognize no significant regularities. . . . And in fact . . . no process based on definite rules can ever manage to generate randomness when there is no randomness before."[82]

Wolfram's assault on the second law is most relevant to the law's application to life and death. Take, for example, fossils—even the fossilized hard parts of unicellular organisms, such as diatomaceous ooze and the White Cliffs of Dover or dome quartz and stromatolites left by filamentous blue-green bacteria. Clearly dead, these fossils should have been randomized if death equaled thermodynamic equilibrium, but their enduring structure stands as a monument to life. It would seem that the life that prepared these fossils defied the second law even after death.

But let us ignore fossils, for the sake of argument. Decaying dead things and corpses bring randomness back into focus. Presumably, denatured viruses decompose, and dead prokaryotes as well as soft-bodied eukaryotes disintegrate under unphysiological conditions or lacking requisite resources for growth. They all would seem to follow the dictates of the second law. Even most corpses of large animals and plants decay and disappear according to the decree of thermodynamics. In none of these cases, however, is one truly considering a living thing. They are all dead things at the point they denature, disintegrate, and decay, and, like other nonliving things, they do not have the power ultimately to defy thermodynamics' second law and avoid the void. But living things have precisely that power!

Corpses and other dead things in their immediate environment (including a variety of decomposers) behave like good closed systems. They reside in the vicinity of thermodynamic equilibrium into which they generally descend.

One may even concede that thermodynamic equilibrium is ultimately the fate of all dead things as it is the fate of nonliving things. But having said that, one has said nothing whatsoever about living things. Indeed, one has said nothing whatsoever about living!

Making Life Safe from Thermodynamics

If life were ruled by the relentless operation of the laws of thermodynamics, life would be driven constantly downward; the development of individuals, their maintenance, and the evolution of species would be virtually impossible; and prolonged life would be sheer fantasy. Of course, one way out of the dilemma is simply to follow François Jacob and deny that life exists at all: "[S]ince the appearance of thermodynamics, the operational value of the concept of life has continually dwindled and its power of abstraction declined. *Biologists no longer study life today.* They no longer attempt to define it. Instead, they investigate the structure of living systems, their functions, their history" (emphasis added).[83] Another way out of the dilemma is to put thermodynamics in its place: in the sphere of closed systems in which it was first, last, and always defined. And in that place, thermodynamics is no longer a threat to life and to the study of life.

Please do not imagine that my quarrel with thermodynamics denies the power of the omniscient and omnipresent second law. Operating in the domain of closed systems close to equilibrium, the laws of thermodynamics have no peer among our human constructions of reality and laws of nature. In their own domain, the laws predict the outcome of actions and reactions correctly and without fail.

But can the second law be invoked to explain behavior in open systems? A familiar example may help one answer the question. How does the second law cope with the financial misadventures of savings and loan banks in the United States or building societies in Europe? Along with owners, including partners or stockholders, depositors, borrowers and creditors, the bank might ordinarily seem like a closed system, running at a profit to be sure but otherwise near equilibrium. Now imagine that the bank's deficits—also called entropy—increase while its assets deteriorate, until, at some point, the bank's assets reach rock bottom, which is to say, their lowest free-energy point—thermodynamic equilibrium—and creditors demand payment for outstanding debt.

The bank faces bankruptcy and closes its door. Depositors are shut out, management panics, and the bank appears hopelessly and irretrievably lost. Nothing would seem capable of returning the bank to an earlier state of solvency. That is exactly what the second law predicts. But what actually happens? In the United States, many a savings and loan bank has been saved from insolvency by the timely intervention of state capital, and many a creditor or stock holder is left holding the bag, while federal insurance rescues the depositors, and management takes home a bonus. In other words, there are exceptions, and exceptions do not necessarily prove the rule.

In fact, the bank never operated as near to thermodynamic equilibrium as one might have been led to believe. The state was there as a buffer between equilibrium and the bank's operation. That buffer sheltered the bank far from equilibrium when the crunch came.

The intervention of state capital that saved the bank does not violate the second law so much as it demonstrates two errors in applying it:

1. The bank as such is not a closed system. The bank is an open system, and its real limits encompass far more than the walls of the bank.
2. The bank as such is not operating near equilibrium. The bank's real distance from equilibrium can only be measured by including state capital and all the laws that protect the bankrupt. Successful business people measure this distance all the time while contemplating their investments.

Similarly, living things are open systems, and they too exist far from equilibrium, although state capital and law are only two among many constraints presently buffering life. A host of institutions prop up life. Indeed, current efforts to save the environment and protect biodiversity depend on state and private capital. But the state and society, the marketplace and supermarket also control our daily life. The twentieth-century social philosopher Michel Foucault is well known for setting this record straight:

In concrete terms, starting in the seventeenth century, this power over life evolved in two basic forms. . . . One of these poles—the first to be formed, it seems—centered on the body as a machine; its disciplining, the optimization of its capabilities, the extortion of its forces, the parallel increase of its usefulness and its docility, its integration into systems of efficient and economic controls, all this was ensured by the procedures of power that characterized the *disciplines: an anatomo-politics of the human body.* The second, formed somewhat later, focused on the species body, the body imbued with the mechanics of life and serving as the basis of the biological processes: propagation, births and mortality, the level of health, life expectancy and longevity, with all the conditions that can cause these to vary. Their supervision was effected through an entire series of interventions and *regulatory controls: a biopolitics of the population.* The disciplines of the body and the regulations of the population constituted the two poles around which the organization of power over life was deployed.[84]

Indeed, the state, society, and society's institutions represent a huge domain between life and the individual as well as between life and thermodynamic equilibrium. The Center for Disease Control, the National Institutes of Health, the Human Genome Project, with all its consortia, the pharmaceutical industry, hospitals, health plans, insurance companies, biotech startups, and virtually every university with a department of biological sciences compete for dollars, for legitimacy and influence. Reproductive technology, anti-aging medicine, exercise and food fads, diets and dietary supplements all broadcast their wares through media, the World Wide Web, around the water fountain and in "smoking areas." Issues such as "Should one have a mammogram before the age of forty-five?" and "Does hormone replacement therapy damage the cardiovascular system?" are debated in Congress and cafes, while the Supreme Court contends with abortion, and the President's Advisory Council makes recommendations on stem-cell research and cloning. One does not merely live and die so much as one is processed, directed, inveigled, and routed throughout life and death. But on no account is one a closed system, and, on no account does one die at the command of entropy!

But in the End . . .

Ultimately, of course, the second law will prevail. "[T]he planet Earth is doomed. The Sun is becoming more luminous every day, and in about 7 billion years its outer atmosphere will have expanded to engulf the Earth. . . . But the physical destruction of the entire Earth is not the only danger. . . . As the luminosity of the Sun increases, the surface of the Earth heats up, making it too hot for life . . . [wiping out life entirely] between 900 million and 1.5 billion years from now."[85] Life, it would seem, has now surpassed something between one-third and three-quarters of its allotted time on Earth. In other words, we too will fall into the pit of entropy, if not anytime soon, then some time in the obscure future. The Earth's apocalyptic ending seems a bit remote, however. In the meantime, it would seem the second law does *not* apply to life and we have only to imagine its consequences somewhere in a future when the previously open system closes in.

Do We Die at the Behest of Chronological Age?

The case for chronology is reinforced by the Bible, which tells of our allotted portion (one hundred and twenty years). But other authorities differ. Shakespeare would have seven ages completed before one is bereft, sans everything, and dies. Since the Enlightment, actuaries, and more recently, gerontologists, have sided with the Bible and raised chronology to the premier determinant of death. Still others feel differently. Salaried employees and wage earners forced to retire at sixty-five may not agree that age is the foremost objective criterion of ability, and if you've ever had a child under twelve years of age at the beginning of the soccer year (September) who is "not eligible to play" as a consequence, you might have other ideas about the objectivity of age to measure ability. Defining biological stages sufficiently to serve as a yardstick for life is difficult, however, because stages are identified primarily by what the organism is doing or how it is making its living, whereas an organism ages chronologically no matter how the organism is surviving or what it is doing. Thus, when it comes to configuring the single most important determinant of death, the balance of opinion falls obstinately to age.

The reason is quite obvious: as a linear variable, age is simply vastly easier to handle mathematically than complex, multifaceted stages of life. Indeed, vital statistics, life tables, mortality tables, life expectancy curves, and survivorship distributions are all built upon a linear model of age. Actuaries make their living predicting longevity as a function of age, and insurance companies thrive by employing age to skew odds toward earnings and profits. Governmental planning takes the age distribution of the population into account, and nonprofit organizations and pharmaceutical companies tailor their products to age-bracketed consumers. But aging, as opposed to chronological age, is not necessarily linear. A variety of traits, from periods of sensitivity[86] to sexual maturity and vulnerability to particular diseases, including "natural death" in old age, are concentrated at certain ages if not narrowly bracketed by age.

LIFE TABLES

Is life really a uniform path to death, or are rate-determining stages hidden in the apparently smooth curves, waiting to be detected in fine-grained analysis? One might hope to find the answer in life tables.[87] Life tables are organized by years and consist of columns and rows of statistics, sometimes combined, for convenience, in groups of years. The title row begins, on the left, with age, usually in years but for close analysis in units as small as days. The remaining columns, separated for females and males, list the computed duration of life remaining in years (or days), known as "residual life expectancy," followed by the probability of surviving another year, known as the "period life expectancy" and corresponding to the likelihood that individuals surviving a given number of years will reach their next birthday.

The problem inherent in using life tables for the purpose of detecting stages of a lifetime is that life tables' statistics are virtually intended to conceal rate-determining steps. The construction of life tables is predicated upon three assumptions: that (1) life follows a mathematical model; (2) mortality is dependent upon an underlying cause; (3) the age-dependency of death is linear. Each of these assumptions tends to smooth out the data and remove distinctions between uniform changes versus rate-determining stages in a lifetime. Thus, discrepancies between law-abiding predictions and data are generally assigned to "biological reality" or error. Furthermore, curve smoothing is aided by the inclusion of additional risk factors inserted into the mathematical model for compiling statistics, including lifestyle and behavior (smoking, imbibing alcoholic beverages, and conduct deemed promiscuous) that alter the calculation of life expectancy, and extrapolation models are added, merging present rates of change in life expectancy with improved medical technology expected in the future.

But one does, indeed, detect humps in life-table statistics. Even the most staunchly devoted follower of life tables concedes that diseases may be clustered within ages: "Prevalence of disease varies with age. Death from cardiovascular disease, for example, is more prevalent among the aged than among the young; the opposite is true for death by infectious diseases."[88] And age-bracketed disease assemblages cause humps in the smoothest curves. We do not, after all, die at the behest of linear, age-centered statistics.

Stages of Life

At the turn of the twentieth century, Karl Pearson attempted to analyze mortality rates for the presence of stages of life.[89] His attempt was heroic, inasmuch as the analysis of curves, even with today's computer-driven methods, remains problematic. Analysis proceeds by deciphering normally distributed data within complex curves. These data, which trace normal curves when plotted individually, have a mean and equal degrees of variability or scatter (the standard deviation) on both sides of the mean. Depending on the type of data, the mean should also correspond (at least roughly) to the mode and the medium or midpoint. The beauty of working with normally distributed data is that the mean can often be attributed to a distinct biological cause, and the scatter around the mean can be attributed to error or chance in the pick of individuals.

Pearson's analysis of mortality curves, thus, began by recognizing bumps and lopsidedness or asymmetry in a curve and proceeded by dissecting (partitioning or parsing out) normally distributed components from the overall curve. Each normally distributed component could then be understood in terms of a biological effect and chance, and a composite curve, reconstructed from the normal curves, could be massaged into something resembling the original curve. Then, the original curve could be interpreted as a product of separate biological effects (i.e., the component biological causes) and chance, on randomly distributed populations.

The bumps Pearson identified in mortality curves for English males[90] led him to resolve five components (i.e., stages) of mortality: "There are five component chance distributions in the resultant mortality curve—five grim marksmen aiming at the throng of human beings crossing the Bridge of Life. However many are the diseases and accidents from which men die, I cannot doubt that they may be substantially classed into five great groups centering round five distinct ages in life." Pearson's components were as follows: "the periods of infancy, of childhood, of youth, of maturity or middle age, and of senility or old age. In the case of each of these periods we see a perfectly regular chance distribution, centering at a given age, and tailing off on either side according to a perfectly clear mathematical law, defined by the total mortality of the period, its standard deviation, and its skewness."[91]

These components may seem familiar and unremarkable, but what was surprising about the conclusion Pearson drew from his analysis was that the scatter around the means drifted far to their right and left. For example, the "mortality of infants" curve was centered at 11 months after birth but stretched from before birth through the ante- or prenatal period and well into the juvenile period. Indeed, according to Pearson's data and curves, in nineteenth-century England, for every 1000 males born, 246 newborns died soon after birth, but 605 would have died in the 9 months prior to birth. Indeed, the sum of ante- and postnatal births accounted for the greatest number of deaths in any of Pearson's components (605 + 246 = 851, compared to 754 deaths in all the other components combined). Pearson's other normal curves also exhibited extensive ranges, reaching far into neighboring stages and beyond.

Pearson's second component was the "mortality of childhood," which killed 46 out of 1000 individuals born. This was the smallest of the normal distributions, indicating a vastly improved likelihood of life. The mean for this curve was at 6 years of age, although the mode (which is to say, the age when the greatest number of individuals died) was in the third year of life.

Pearson's third component was "mortality of youth," during which 51 young adults died with a midpoint at 23 years and a range extending from birth to nearly 45 years. The fourth component was "middle age mortality," during which 173 individuals died with a mode and mean of about 42 years and a range extending from five to sixty-five years.[92] Finally, in the fifth component of "old age mortality," 484 males died with a mean of 67 years, a mode of 72 years, and a range from under 20 years to 106 years.

Pearson did not succeed in his attempt to find unique biological effects responsible for deaths in his five components. He began by waffling about "infant mortality," attributing these early deaths vaguely to "developmental failures," which, as a stanch eugenicist, he laid at the feet of "bad parentage." And he was no more trenchant in his attribution of killers in the four other components. Raymond Pearl complained that "brilliant and picturesque as is Pearson's conception of the five Deaths, actually there is no slightest reason to suppose that it represents any *biological* reality, save in the one respect that his curve fitting demonstrates, as any other equally successful would, that deaths do not occur chaotically in respect of age."[93]

Indeed, Pearson's components are under or ill defined, and Pearson's dismissal of linkages between certain diseases and particular age groups, "with the possible exception of childhood,"[94] seems hasty. Hence, Pearson's stages are not widely cited. In fact, only one of his components is alluded to by name, namely, old age mortality. In his words, "what we have termed old age mortality is only that special group of causes most active in old age, that group which in England carries off nearly one-half of human beings. This group corresponds in some sense to the natural end of life, but this natural end may come long before old age."[95]

"The natural end of life" peaks in old age, but covers a large range. It is a stage but not a disease, as such. Rather, the "causes grouped together in [this component do not] . . . refer so much to the special severity of certain diseases, as to the special prevalence during the period considered of various susceptibilities, relative capacity to resist, or it may be incapacity to resist death."[96]

Pearson may have to shoulder responsibility for the notion of a natural death at old age, but he very clearly had in mind the possibilities of equally natural deaths at other ages. And if Pearson is to be saddled with the notion of multiple periods of natural death—recurrent series of episodes, components, or stages of death—then he is to be credited with conceiving of death as moving on through multiple, multifaceted, and multidimensional stages.

Huge ranges around the means characterizing Pearson's five components highlight the problem of identifying stages. Certainly, populations of slightly different individuals and life's continuity lead inevitably to huge overlaps. Indeed, Pearson's components defy unique criteria including groups of diseases, and intermediate stages are easily set between the components. Nevertheless, vigorous or not, overlapping or not, reasonable criteria permit the division of lifetimes into recognizable rate-determining stages.

Of all the stages of life, death at the earliest stages of development, or early pregnancy loss (EPL)—typically within six to seven weeks of gestation—is shockingly high. Fertilized eggs fail to cleave, pre-embryos or blastocysts fail to implant, and embryos and fetuses die *in utero* and are resorbed or aborted. "According to recent data, [death] is the fate of 70–80% of fertilized human egg cells, and in the majority of cases the end comes at such an early stage that nobody usually notices it."[97]

Indeed, "the concentration of anomalies among spontaneous abortions is about a hundred times that at birth. About 95 percent of chromosomal anomalies abort. Trisomy 21 [an individual with three chromosomes 21] is more viable than most, and even of these about 70 percent abort."[98] In the United States during 1999, of 6.23 million registered pregnancies, 1.0 million (16 percent) were lost as fetuses, *excluding* induced abortions.[99] And like trisomy 21, EPL is correlated with age of the egg.[100] On the other hand, in vitro fertilization with eggs from young donors (<35 years), indicate that EPL is not correlated with the age of the mother (that is, with uterine age[101]).

The Seven Stages of Life

Within (or without, depending on point of view) broad limits, researchers seem to accept seven age-centered stages following birth, loosely corresponding with the "seven ages of man":

1. neonate (first year; sometimes broken down into ante- and postnatal, corresponding to birth to 28 days, and 29 to 364 days)
2. infant/toddler (1 year through 4 years)
3. juvenile (5 to 14 years [in the U.S. the age of presumptive puberty is 12 years for girls and 14 years for boys])
4. adolescent/pubescent (15 to 19 years [frequently extended to 24 for convenience])
5. young adult (20 [or 25] to 44 years [generally broken into two periods: 20 or 25–34 and 35–44])
6. older adult (45 to 74 years [generally broken up into two or three periods of 45–54; 55–64; 65–74])
7. senescent adult (over 75; generally broken down into 75–84 years and 85 years and over)

As Pearson might have forewarned, proneness to discrete causes of death cannot be attributed to each stage with any confidence. Nevertheless, accidents account for a disproportionately large part of deaths occurring during the first four stages. Adolescents and young adults experiment with life more as a result of curiosity than wisdom and suffer a disproportionately high mortality from suicides and homicides. Older adults are more settled but are increasingly vulnerable to the ravages of disease. Death rates, while remaining high, do not increase among senescent adults.

And particular classes of disease are clustered in the different stages. Partial disease profiles (see figure 1.2) show the proportion of deaths due to the leading classes of disease in human beings in age-bracketed groups.[102] Each group is seen to be associated with a unique mix of killer diseases. Diseases of the heart and cerebrovascular diseases, for example, kill a far greater percentage of neonates and senescent adults than toddlers, juveniles, pubescents, young adults, and old adults. On the other hand, malignant neoplasms are the major killers of toddlers, juveniles, pubescents, and young and old adults. This conclusion is surprising, since cancer is so often considered a disease of the elderly. Moreover, the proportion of tumors switches following age 10 from predominantly hematopoietic, nerve, connective tissue, and epithelial to more than 90 percent epithelial after age 45 years. Chronic lower respiratory disease drops off in young adults, picks up again, and peaks in senescence.

One can easily think of many reasons that diseases should precipitate out at particular stages: diseases typically kill by disrupting bodily functions, systems, and organs that operate predominantly at particular ages; an agent or agents that deliver particular diseases may have age-specific requirements in their host, and hosts may be susceptible during limited periods; diseases may be associated with incubation periods before becoming conspicuously debilitating and causing an individual's death. Thus an age configuration enters the calculus for disease in several ways: periods of vulnerability, development, duration, onset and severity of symptoms, and maturity, all of which would seem to circumscribe physiological stages, as opposed to mere chronological age.

Diseases caused by some parasites and pathogens require a vector for their delivery that may determine when an individual can be exposed. Something rudimentary to the organism, rather than age as such, may be crucial. Typically, the difference between a parasite and pathogen is size, pathogens being the smaller and taking smaller bites at a time, while parasites take larger chunks. Pathogens include viruses, bacteria, and mold, while parasites include protozoa and animals (for example, trypanosomes, cestodes, flukes). All that these agents have in mind, so to speak, is living off our bodies, but, in consequence of their way of life, they risk killing us in the process and losing their meal ticket. Host size, and hence, stage of development, may be the deciding factor in whether one resists an attack or succumbs to it.

Some pathogens and parasites kill more or less directly through anorexia, diarrhea, and weight loss, while others kill indirectly. Some produce progressive symptoms, such as mental deterioration, stooped posture, seizures, and paralysis before death, while an intestinal parasite may kill through anemia. Death due to parasites and pathogens may, thus, be stage-typical if not entirely stage-dependent.

Some parasites, pathogens, and genes (for example, Huntington's disease) may be tolerated more or less well by the young or the diseases they cause may seem to have long incubation periods, confining them to advanced age. For example, individuals with various forms of mammalian spongiform encephalopathy caused by a slowly emergent prion—a pathological configuration of a protein that is normally found in a healthy (physiological) configuration—only become symptomatic with advancing age. Similarly, diseases associated with dementias, neuronal degeneration, and progressive mental deterioration—memory loss, confusion, and disorientation—are frequently associated with excessive intracellular neurofibrillary material that appears tangled and with plaques of granular or filamentous masses. In these cases, the age of onset may conceal a stage-specific vulnerability. Breast cancer also takes a high toll of women when it strikes late in life, after the reproductive period is over.

Like old age, stages of a lifetime do not seem to be the killers they might seem: we live despite them rather than die because of them. In fact, we live for

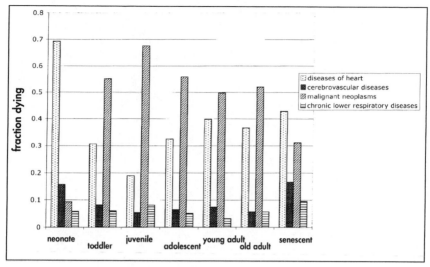

FIGURE 1.2. Partial disease profile for the seven stages of a lifetime. (Data from Centers for Disease Control and Prevention, 2003, tables 36–41.)

a lifetime through a variety of stage-typical causes of death, although most of us will not live through all of them.

Ultimately, death must be redefined in terms of life's stages. Death must be understood as complex, involving everything about an organism, its anatomy and physiology, its interactions with its environment, and its inherent genetic potential and limitations.

IN SUM

The attempt to demystify death in chapter 1 is a necessary preliminary to death's analysis. The problem of understanding death is not solved by scientific and technical vagaries but by hardheaded research and hardnosed fact. Chapter 1 has shown that life span, aging, and dying are not adapted to making room for variants, not determined by genes, not required by the environment, and not decreed by entropy.

Death is shaped by life. Survivorship distributions illustrate the sensitivity of aging, dying, and death to both randomness—accidents and encounters with environmental hazards—and determinism—biologically built-in and genetic effectors. Most remarkably, genes encoding the proteins constituting age-related pathways are similar (that is, conserved) in organisms as distantly related as yeast, roundworms, flies, mice, and human beings, but death also comes in different forms in these same organisms. In this framework of complex interactions, of metabolites, genes, pathways, cells, tissues, organs, organisms, and environments, aging, dying, and death, like life itself, would seem to be subjects of intense and practical negotiation and compromise. The chapter leaves little doubt that death will come in different forms as life moves through its different stages.

Chapter 2

Charting Death's Evolution and Life's Extension

Past progress against mortality may be underestimated, and as a consequence, predictions of future progress against mortality may be too low.
—J. W. Vaupel, K .G. Manton, and E. Stallard,
"The Impact of Heterogeneity in Individual
Frailty on the Dynamics of Mortality"

Thus, we know that the upper tail of the age distribution of deaths has been moving steadily higher for more than a century.
—J. R. Wilmoth and H. Lundström,
"Extreme Longevity in Five Countries"

He was beginning to understand that death was not simply a loss of vitality, but a profound change . . . [that] made him fear death less. You did not suffer after death. There was nothing left to suffer.
—Paul McAuley, *Pasquales's Angel*

Death's evolution? What irony! How could death evolve when it has no apparent benefit for the survival or reproduction of individuals? How could the manufacture of corpses, aka death, evolve? What would possibly constitute evidence for death's evolution?

Intuitively, one might suppose that the evolution of death would make life shorter and faster, for instance, via progeria disease. But if death, like everything else in life, evolves through the propagation of life, then death could hardly evolve through early-onset diseases, especially where there is no off-setting balancing mechanism or even progeny to carry on the trait. And if death, like so much else in life, represents an adaptation, somehow death evolves by becoming more efficient—by allowing organisms to die only when all other alternatives have failed. *Thus, counter to intuition, death would have to evolve by forestalling the metamorphosis of organisms into corpses!*

A physical metaphor might help clarify the idea of death evolving by pro-longing life. Remember (or don't remember if it's too painful) your introductory physics lab (or imagine one if you never took physics with lab): You rolled a marble down an inclined plane (actually a grooved track) and measured how far the marble rolled after reaching a level plane (actually a table top). Had the instructor permitted, you might then have rolled two identical marbles down two identical tracks, except marble #1 rolled on a smooth track and marble #2 on a rough track. The force of gravity accelerating the two marbles would be the same, but marble #1 would encounter less friction (resistance) and travel farther than marble #2 on the level plane.

Now imagine two lives rolling down two lifecycles, life #1 rolling on a well-lubricated lifecycle, and life #2 rolling on a poorly lubricated lifecycle. Just as marble #1 rolled for a greater distance than marble #2, life #1 lasts longer than life #2, that is, the organism on the well-lubricated lifecycle lives longer than the organism on the poorly lubricated lifecycle. Let us say, by way of illustration, that the lifetime on the well-lubricated track turns out to be twice as long as the lifetime on the poorly lubricated track. For example, if life #1 travels two units of life while life #2 travels only one unit, life #1 will be twice as long as life #2. Moreover, the corollary is also true: life on the poorly lubricated lifecycle (traveling one lifetime per unit of life) proceeds at twice the rate as life on the well-lubricated lifecycle (one lifetime per two units of life). This is only to say that *longer life proceeds more slowly!*

What it boils down to in terms of Darwinian fitness is quite straightforward if counterintuitive: all else being equal, *where death is evolving, the duration of lifetimes is increasing and the rate of living is decreasing.* Thus, longer life and slower living are the qualities that one would look for were death evolving.

MEASURING DEATH'S EVOLUTION: EMPIRICAL EVIDENCE

Death's competitive advantage is readily traced in *Homo sapiens,* since a great deal is known about human longevity. The fact is, we are big winners in the race toward longer and slower lifetimes! Contemporary human beings have a remarkably long lifetime—with a median of nearly eighty years or thereabouts in developed countries. Indeed, with the possible exception of the bowhead whale, human beings are the longest living mammals on Earth. In contrast, our ancient reptilian cousins, the dinosaurs, seem to have died young: bones from the oldest-known *Tyrannosaurus rex* (Field Museum, FMNH PR2081) indicate that it died as a senescent adult at an age of only twenty-nine years.[1]

But did our species emerge with its extraordinary lifetime or did it evolve that way? The answer from empirical evidence is unambiguous and undeniable: lifetime is evolving upward and its reciprocal, mortality, is evolving downward.[2] We are living longer! Indeed, we are living in an age of rampant life extension: our lifetime—the duration of an individual's life from birth to death—is increasing, and human beings are surviving to older ages and doing it in better health than ever before.

Of course, as we evolve, the things that kill us also evolve, and, in the eternal "arms race" between the killed and killers, one sometimes wonders who is winning. Frankly, we are losing in the battle against the rampant spread of resistant forms of malaria that are devastating children in subtropical and tropical regions of the world, of methicillin-resistant (and now vancomycin-resistant) flesh-eating *Staphylococcus aureus,* and TB, to say nothing of HIV infections spreading in newborns and young adults. I also cannot discount 3.7 billion people—more than half the population of *Homo sapiens*—presently suffering from malnutrition, the largest number in history according to the World Health Organization, and I do not concede to progress all the so-called accidents, wars, greenhouse gas emissions, and massive new sources of other pollutants and hazards (especially in the kitchen and bathroom) threatening modern life.[3] But caveats (or forecasts) notwithstanding, the good news is that each of us in the developed and developing parts of the world can reasonably expect to outlive our parents and enjoy the fruits of our labor to a great old age.

Comparisons with previous generations are easily made with the help of abundant data on lifetime and survivability engraved on headstones, inscribed in registries, and revealed by the analysis of bones from long-forgotten ancestors. Although not of equal quality and accessibility, these data can be analyzed and compared after computing statistics on rates or probabilities of change, specifically (1) mortality or death rates (aka "force of mortality") computed from the number of individuals (component units) of a group exposed to the risk of dying who expire in a given interval; and (2) life expectancies or the years people born in a particular year (a cohort) are most likely (or can expect) to live after attaining x years ("life-after x"). Overall, mortality rates tend to decrease and life expectancy to increase, if not quite globally or uniformly, at least, more or less, consistently.[4]

Indeed: "For the males the crude data analysed here indicate a decline in mortality starting in the Early Middle Ages [fifth century], during or after the fall of the Roman Empire. This late decline of male mortality is paralleled by the females. The decline is even more pronounced among females than among males. The increase of female mortality leading to the decline of survival from the Mesolithic to the Neolithic is probably a very important shift in the selectional forces shaping our species."[5] Thus, life expectancy has been increasing for fifteen hundred years, or thereabouts, and throughout Western Europe. In France, "life expectancy at birth went from 24 years in 1745 to 42 years in 1850 for males and went from 26 to 43 years for females."[6] While in the United States "life expectancy at birth has increased from 47 years in 1900 to about 75 years in 1988."[7]

And the trends go on. Data from the National Vital Statistics Report of September 2001 demonstrate generally declining death rates by age for males and females in the United States from 1955 through 1999 (figure 2.1). Precipitous declines occur among infants (under 1 year) and juveniles (1–4 years and 5–7 years), although death rates for young adults (15–24 years and 25–34 years) and the elderly (86 years and over) are not depressed nearly as conspicuously.

Even within one year (1998 to 1999, figure 2.2), the percentage change in death rates is seen to drop in most age categories (with the exception of young adult women [15–24 years] and the elderly 75 years and over), although much of the drop disappears when data "[i]ncludes races other than white and black." And these overall trends in death rates are expected to continue.[8]

Given the robust increase in population size experienced throughout the world, increased numbers of individuals living into the early and middle ages

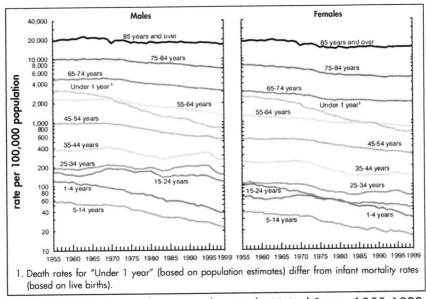

1. Death rates for "Under 1 year" (based on population estimates) differ from infant mortality rates (based on live births).

FIGURE 2.1. Death rates by age and sex in the United States, 1955–1999. (From *National Vital Statistics Report*, September 21, 2001, p. 7).

FIGURE 2.2. Percentage change in death rates and age-adjusted death rates between 1998 and 1999 by age, race, and sex in the United States. (From *National Vital Statistics Report*, September 21, 2001, p. 5). (Age-adjusted death rates based on year 2000 U.S. standard population.)

Age	All races[1]			White			Black		
	Both sexes	Male	Female	Both sexes	Male	Female	Both sexes	Male	Female
				Percent change					
All ages:									
Crude	1.4	0.6	2.2	1.5	0.8	2.3	1.1	0.3	2.0
Age–adjusted	0.7	-0.3	1.5	0.7	-0.3	1.5	1.0	0.1	1.8
Under 1 year[2]	-2.6	-2.0	-3.4	-3.8	-2.3	-5.5	-0.3	-1.4	0.9
1–4 years	0.3	2.4	-1.9	2.0	4.3	-0.4	-4.4	-4.6	-4.5
5–14 years	-3.5	-5.1	-0.6	-2.7	-4.2	-0.7	-2.4	-2.8	-2.2
15–24 years	-1.3	-2.8	2.8	-1.2	-2.5	2.4	-2.7	-4.6	-3.6
25–34 years	-1.2	-1.0	-1.8	0.2	0.4	-0.2	-5.3	-4.8	-6.2
35–44 years	-0.2	-0.7	0.7	0.0	-0.5	1.1	-1.5	-2.0	-0.8
45–54 years	-0.9	0.7	1.1	1.1	0.9	1.3	-0.1	-0.1	-0.2
55–64 years	-0.9	-1.3	-0.2	-0.8	-1.3	-0.2	-1.2	-1.1	-1.3
65–74 years	-0.4	-1.1	0.3	-0.6	-1.3	0.2	0.2	-0.1	0.3
75–84 years	0.8	-0.3	1.7	0.8	-0.3	1.6	2.0	0.5	3.2
85 years and over	2.4	1.0	3.0	2.4	0.9	3.0	4.5	3.3	5.0

1. Includes races other than white and black.
2. Death rates for "Under 1 year" (based on population estimates) differ from infant mortality rates (based on live births).

of their lifetime might be expected to turn into more individuals reaching old age, and more old individuals might become more centenarians (100 to 109 years) and supercentenarians (110 years and more) in turn. Thus, more people would enter the ranks of those approaching the asymptote late in life simply because there are more people. Indeed, in "developed countries the number of people celebrating their 100th birthday multiplied several fold from 1875 to 1950 and doubled each decade since 1950."[9]

But, overall, the increase in life expectancy is not a statistical artifact. In fact, statistics has very little to do with it: in two populations differing in size by a factor of ten, the oldest individual in the larger population is most likely to be only one year older than the oldest individual in the smaller population.[10] Better treatment of infectious disease among the young seems to have more to do with the prolongation of longevity than population size.[11]

Figure 2.3 shows the life expectancies at birth for people born between 1970 and 1999 (computed from census data compiled in the National Vital Statistics Report of September 21, 2001, corrected for the 2000 census). Sorted by gender and by white and black self-identified "races," these values show more or less parallel increases despite large gaps between the curves and some conspicuous ripples. Differences in life expectancies between individuals grouped by race are generally attributed to local inequities and social forces, although responsibility for the extraordinary dip in the curve for black males between 1985 and 1995 may be laid at the door of those who poured crack/cocaine into the black community during this period.[12] Hopefully, same-gender differences between races will disappear with the expansion of opportunity and consequent improvements in standards of living.[13]

Difference in data for females and males are not easily attributed to circumstance and environments, however, since men and women tend to live together.[14] These differences in the lifetimes of females and males are widespread, across the board of ages, and are, if anything, increasing. In "most developed countries . . . average death rates at ages above 80 have declined at a rate of 1 to 2 percent per year for females and 0.5 to 1.5 percent per year for males since the 1960s"[15] (see figures on total deaths and infant mortality in Singapore in chapter 4). In the United States, "the age corresponding to the survival of 0.001% of new-born children" increased 6 percent for men and 8 percent for women between 1900 and 1980.[16] In Japan, the differences between life expectancy at birth in males and females increased from 3.40 years in 1950–1952 to 6.97 years in 2003.[17] Women would appear to lead the trend in life extension for biological reasons, although what these reasons are remains shrouded.

The widespread belief that natural selection acts slowly rather than rapidly seems to raise an impenetrable barrier to the notion that the rapid changes in the duration of a lifetime are due to evolution. But a belief is only a belief. Backed up by neo-Darwinism and the "modern synthesis," natural selection is

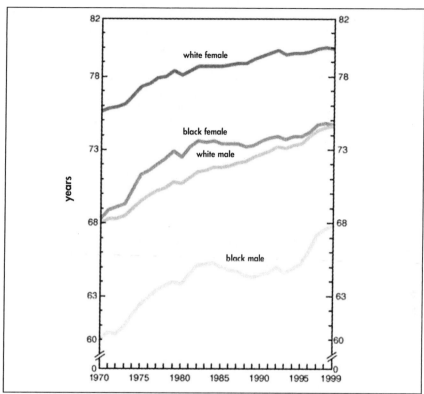

FIGURE 2.3. Life expectancy by sex in the United States, 1970–1999. (Data from *National Vital Statistics Report*, September 21, 2001, corrected for the 2000 census).

sometimes equated to the gradual accumulation of miniscule quantitative changes attributed to minor mutations, but many of those taking a closer look at the rate of evolution in the wild find evidence for evolution proceeding rapidly, if only intermittently. Certainly, several valid mechanisms for evolution, from drift to advantageous mutations, bottlenecks and colonizers, suggest that evolution may proceed at different rates, some faster than others.[18]

The point is that contemporary human culture is not changing slowly, and human evolution is currently being shaped by human culture. From the perspective of niche-construction, "that the activities of organisms bring about changes in their environments,"[19] one would hardly be surprised that lifetime and death are evolving rapidly. By any calculation, those benefiting most by any or all the devices currently prolonging life—from urbanization and improved sanitation to adequate nutrition and advanced medical care—are those whose fitness is most increased by those devices. In other words, *the*

environmental changes currently in play and prolonging individual life may
very well have snowballed into evolutionary mechanisms that delay death.

MODELING THE EVOLUTION OF LIFETIMES

The notion of death evolving by lengthening the lifetime is the kind of seemingly absurd notion that every scientist hopes to conjure up at least once in a
career, because if the absurdity turns out to be demonstrably plausible, it is not
as absurd as it first appears and may even become a curiosity worthy of empirical investigation. Fortunately, the plausibility of the absurdity at hand is easily
demonstrated with the help of a model, and, as it turns out, death may very
well evolve by lengthening life, in particular, by prolonging into adulthood the
well-being and vitality intrinsic to the juvenile stage.

The model is adapted from the stochastic chronic disease model of survival.[20] In effect, the life span model treats life as a chronic disease we pass
through in stages. Typically, chronic diseases are nervous and mental diseases,
other diseases of the central nervous system, heart diseases, all forms of tuberculosis, hardening of the arteries and high blood pressure, diabetes and many
forms of cancer. "Generally, chronic diseases advance with time from mild
through intermediate stages to severe stages to death. Often, the process is
irreversible but a patient may die while being in any one of the stages. Patients
in a given stage of a disease not only are subject to different forces of mortality than those in another stage, but also may advance to the next stage and
experience a greater chance of dying. The process is dynamic, and the stages
of the disease are a dominating factor in the survival or death of a patient."[21]
In other words, chronic diseases are the kind you lose work and sleep over as
they become progressively debilitating. And as you become increasingly inca-
pacitated, you become more and more vulnerable or fragile until you die.

That's also true of a life span generally. Thus, adapting the chronic disease model to a life span simply requires transforming the progressive phases
of disease to rate-determining stages of life. The model then allows one to
examine the consequences of expanding the life span by altering the values
assigned to parameters governing all or any of these stages.

One is accustomed to thinking about death at the conclusion of a disease
or old age, but only some sufferers of a chronic disease will reach the severe
or terminal stage, and only some of us will live through the ripening of senescence before death strikes at extreme old age. In fact, "the idea that variability
in lifetimes is largely determined by a multi-stage process of destruction not
only agrees with the well-known facts about the heterogeneity of the population with respect to the death risk, but also permits an explanation of the possible reasons why this heterogeneity might arise."[22] Much as one may die in

any phase of a chronic disease, one may die in any stage of a life span, and much as one may live through any phase of a chronic disease except the terminal one, one may advance through any stage of a life span except the terminal one.

The stages of a life span should not be thought of as arbitrary, age-bracketed segments possessing no identifying characteristics of their own. Although the stages fall in line progressively and the age-specific probabilities of death change continuously, some stages are notoriously difficult to negotiate (for example, gastrulation) and the transitions between some stages (prepubescent to adolescent) represent crucial nodes or points of inflection in a life span. Furthermore, although some stages may slide easily into one another (young adult to old adult if one escapes a midlife crisis), other stages represent discordant assemblages of causes of mortality (see figure 1.2).

In the present model, stages of the life span from embryo to senescent adult have a specific duration, and end when the individual dies or progresses to the next stage. The time an individual spends in a stage may not approach the full duration of that stage, since the duration of any stage may be reduced by death and moving to the next stage follows a distribution rather than a limit. Indeed, the average duration of stages for individuals in a population, at any time, will fall short of any maximal duration (if there is a maximum). The duration of each of the stages is, nevertheless, a species-typical characteristic.[23]

Life as a chronic disease (figure 2.4) begins with the embryo and fetus and ends with old adult and senescent adults.[24] The overall trajectory is toward death (diagonal arrows), but individuals move at each stage to the next stage as well (vertically descending arrows). Turning the figure into a model merely requires the addition of functions for the different rates of movement from stage to stage and stage to death. The number of individuals surviving at any stage (x) is deduced from life tables (by bracketing off groups of cohorts according to age).[25] The number of individuals surviving at any stage provides the data for a probability distribution among stages. The arrows illustrate the direction of transition, either from one stage to another or from a stage to death. The force of transition from stage to stage (equivalent to the force of morbidity or incidence rate in the chronic disease model) is represented by $\cdot V(x)$ for each stage, and the force of mortality is $\mu(x)$ (or intensity function) for each stage. These rates, or the intensity of transitions, are functions of the stages.

The pies and segments in figure 2.5 illustrate running the chronic disease model through the eight stages of a life span, proceeding in a curve (open arrow) from lower left to lower right. Segments of the pies with changing tones of gray represent the stages (blocks in figure 2.4 with corresponding shades). The size of these segments in each pie reflects the portion of the cohort still alive, while the change in the size of these segments is due to the

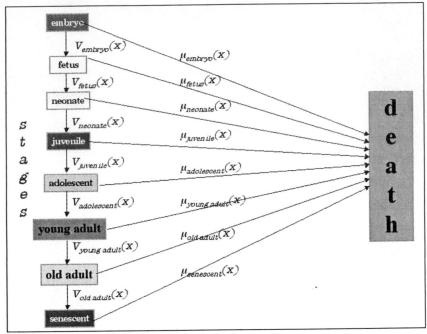

FIGURE 2.4. Life span model.

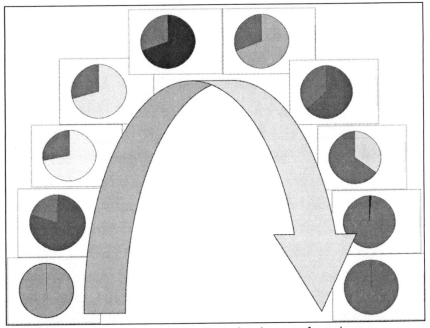

FIGURE 2.5. Life span model in action. The demise of a cohort.

rate at which the cohort moves through a stage $(V[x])$ and dies at each stage $(\mu[x])$. The segments decreasing in size (having different tones of gray) represent the portion of the cohort surviving from one stage to the next; the segments increasing in size (constant middle tone of gray) represent the portion of the cohort dying out until the entire cohort has expired (last pie).

ACCOMMODATING INCREASED LONGEVITY

How can this model of movement through life's stages and toward death be adapted for simulating the evolution of lifetime extension? One can imagine a life span being extended in either of two ways: overall lengthening versus lengthening in specific stages, that is, either by lowering the death rate throughout or by expanding one or more stages with below-average death rates.[26] The question is, does the life span lengthen as a whole, like an accordion, or does the most longevous stage expand while other stages remain the same, like a bagpipe?

UNIFORM LIFETIME EXPANSION: THE ACCORDION MODEL

In the accordion model, figure 2.6, the duration of each stage expands more than it contracts as members of the cohort die. An increase in life span thus reflects net expansion, that is, an overall expansion minus contraction.

The expansion of stages is easily simulated by decreasing the rates at which the cohort moves through the stages $(V[x])$. The scheduled rates of dying $(\mu[x])$ are unchanged. The result of running the model with expanded stages is shown in the figure. The probability of dying is reflected in the size of the death segment (constant middle tone of gray), while the probability of having reached every stage corresponds to the size of the other segments (with tones of gray corresponding to blocks in figure 2.4).

Thus, the accordion begins with a cohort of preembryos (lower left). Additional segments of the accordion, representing embryos and organisms that have died (intermediate gray), are added to the segment representing the preembryos preserved in part as a consequence of expansion. Similarly, in succeeding iterations, segments are added as prior segments remain patent due to expansion: fetuses are added to embryos; neonates are added to fetuses; juveniles are added to neonates; adolescents are added to juveniles; young adults are added to adolescents; old adults are added to young adults; and senescent adults are added to old adults, while, at the same time, the size of segments representing deaths increases but does not take over as it does in figure 2.5.

Death does not catch up with the expanded stages! The unexpected consequence of uniform expansion is that the expanded accordion does *not* contract

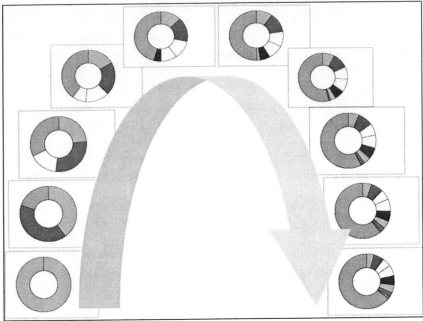

FIGURE 2.6. The accordion model. Uniform expansion of a life span.

completely. Neither the progression through stages nor the intrinsic rates of death at any stage picks up the slack. "Holdouts," thus, remain at every stage.

The dynamics of the accordion model expose its inherent weakness. Although each stage contracts as some members of the cohort move to the next stage and some die, all the individuals in any stage are not eliminated with each iteration. The consequences are bizarre: at the same time that some members of the cohort enter senescence, others are still embryos, fetuses, neonates, juveniles, adolescents, young adults, and old adults!

Thus, the accordion model leads to the unacceptable prediction that representatives of each stage of the life span linger. Of course, stochastic drift would ultimately remove all members of the cohort at every stage, but the idea of a prolonged embryo or fetus, and hence, of prolonged pregnancy, violates well-known restraints on the duration of pregnancy. One expects development in the embryo and fetus to be exquisitely timed and not lengthened or shortened. Even pregnancies reaching three weeks beyond the expected due date are likely to have fatal consequences for the fetus.[27] Predictions forecast by the accordion model are, therefore, rejected, and life spans are not expected to expand uniformly.

STAGE-LIMITED LIFETIME EXPANSION: THE BAGPIPE MODEL

Like a bagpipe with a constantly puffed-up bag exciting steady base tones and permitting a resonating treble melody from pipes, the bagpipe model simulates increased longevity by expanding at a stage with a lower-than-average intrinsic death rate. That stage is the juvenile stage, which falls into a trough in the overall mortality curve between the higher death rates for neonates and adolescents (see chapter 4). Indeed, the low death rates of juveniles has been consistently reported by the Centers for Disease Control and Prevention (CDC) even when new 2002 statistics show an increase in infant mortality rates (from 6.8 deaths per thousand to 7.0 deaths per thousand—the first increase since 1958).[28] An increase in lifetime would thus depend on the extension of the low juvenile stage's death rate into subsequent stages.

The dynamics of the bagpipe model are illustrated in figure 2.7. The expansion of the juvenile stage is simulated by equating the rate of individual deaths in later stages $(\mu[x])$ to the lower rate of the juvenile stage, while retaining the rates at which stages move ahead $(V[x])$. At each stage, a wedge-

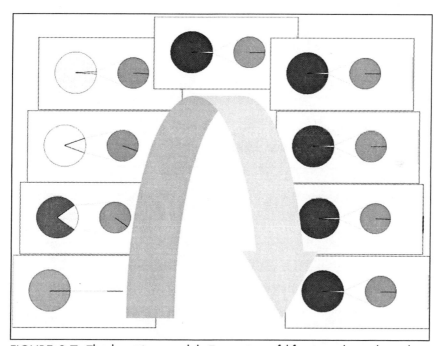

FIGURE 2.7. The bagpipe model. Extenstion of life span through prolongation of juvenile stage.

shaped segment of the pie represents individuals dying at that stage. This segment is removed from the cohort and added to the dead part of the cohort (the small disk to the right). The radial line drawn on this disk indicates the rate of death in the cohort; it approaches the horizontal (slope 0) at the juvenile stage and remains there as members of later stages continue to die at the rate of juveniles.

Thus, the bagpipe model begins with a cohort of preembryos (lower left). At the first iteration, a portion of the preembryos has moved to embryos, while another portion has died and moved out of the cohort (to the small disk) at a high rate (angle of radius). At the next iteration, a portion of embryos has moved to fetuses, while another portion has died and moved out of the cohort (to the small disk), also at a relatively high rate (angle of radius). Similarly, in the succeeding iteration, fetuses have transited to neonates, while yet another portion has died and moved out of the cohort (to the small disk), now at a slower rate (angle of radius). The same phenomenon repeats itself at the next iteration with the neonate's transitions to juvenile, and the death rate reaches its minimum (radius approaches horizontal). This last transition marks the turning point in the illustration, however, since—although individuals continue to progress through the stages—the death rate has remained at a minimum corresponding to that of the juvenile stage (hence the disk is virtually unchanged, and the radius continues to approach the horizontal).

The bagpipe model thus simulates lengthening the lifetime by spreading the juvenile's low mortality rate to post-juvenile stages. At the same time, the spread of the juvenile life expectancy to adult stages implies life's slowing down. Indeed, as adults remain youthful, life would seem to continue all but unchanged.

Supporting data are abundant. For example, as "[c]hildhood mortality has decreased by 90%," adult life expectancy has increased by 50 percent over the last 150 years in some countries in northern Europe.[29] This upward spread of lifetime extension has even surprised sophisticated gerontologists. Caleb Finch and Eileen Crimmins observe, "[t]he rapid decline in old-age mortality in developed countries that began in the latter part of the 20th century took researchers by surprise . . . [b]ecause few medical breakthroughs occurred before the onset of the decreased mortality from heart disease and stroke in the 1960s."[30] In effect, mortality curves retain the lower values of youth while cohorts grow chronologically older. Thus, *we stay younger and age more slowly while growing older!* The question that remains is how death's evolution slows down life and allows the resistance and robustness associated with our juvenile stage to seep into our adult life.

IN SUM

Death's evolution would seem paradoxical: how could producing corpses possibly be adaptive to life? The paradox disappears, however, when death is seen to evolve by lubricating the lifecycle, which is to say, by promoting slower living and delayed dying. Hence, death evolves through life extension.

In fact, we are living longer today than at any time in the past, and the prolongevous change in *Homo sapiens* is precisely what seems to be happening during recorded history. Indeed, life extension is evident in data garnered over centuries and years, from the bones of our ancestors to contemporary registries of births and deaths.

The chronic disease model of survival, adapted to the life span, suggests that the average human life span has not expanded uniformly (the accordion model), since uniformity would lead to developmental imbalance and an unrealistic distortion in the duration of stages. But life's overall expansion is compatible with the prolongation of the juvenile stage (the bagpipe model). We seem to be living longer because we are being juvenilized over our adult life.

Chapter 3

Rethinking Lifecycles and Arrows

There is no generally valid, orderly relationship between the average duration of life of the individuals composing a species and any other broad fact
now known in their life history, or their structure, or their physiology.
　　　　　　　　　　　　—Raymond Pearl, *The Biology of Death*

He knew it was impossible for human physiology to change at less than
glacial speed, but he suspected that some shocking transformation had nevertheless taken place in what was required to sustain human life.
　　　　　　　　　　　　—Donna Leon, *Willful Behaviour*

Even were one to concede that death is part of life, one would want to see how
death is integrated into life's other features before accepting death as evolving.
Well, let us not mince words: one must reorient oneself entirely to life in order
to integrate death.

　　I cannot pretend that biologists share a single view of death's integration
into life. In fact, the two most widespread views—the cyclic view and the
linear arrow view—portray life, and hence death, in entirely different ways.
When life is viewed as a cycle, death is seen as the continuous ejection of
corpses, while, when life is viewed linearly, death is seen as the endpoint for
living things (bodies or somas, thalli, clones, and cells), branching off of an
immortal germ stream. In cycling life, death plays an active role, and, thus,
death has the potential to evolve. On the other hand, in linear life, death is
excluded from an active role in shaping life and therefore the possibility of
death's evolution is barred. Thus death can evolve only if life is cyclic.

LIFE AS A CYCLE: LIFECYCLES CONNECT LIFE TO LIFE

The notion of life cycling is really not that old, having been formalized as biogenesis by Thomas Huxley at the fin de siècle in contrast to the defunct notion of spontaneous generation, or life arising spontaneously from nonliving matter. Huxley's doctrine was immediately distorted into a new version of William Harvey's "all life comes from eggs," and Rudolf Virchow's "all cells come from cells" and corrupted into Ernst Haeckel's "biogenic law," that all life repeats its evolutionary history. Today, reinforced by molecular cladistics, biogenesis has returned to Huxley's original doctrine that *life is the source of life, and, hence, life cycles* (open arrows in figure 3.1).

Since the late nineteenth century, a "life cycle" or "life-cycle" has stood for life's continuity.[1] Here, "lifecycle" (the equivalent of "life's cycle" but without a space, hyphen, or apostrophe) is employed with the intention of signifying continuity among life's molecules, its cells, and organisms. This continuity also connotes an energetic relationship.

In the ideal, Newtonian world, cycles are endless, and simple harmonic motion goes on forever. But, in the real, macroscopic world of living things, nothing moving fails to encounter resistance, and, hence, nothing moves forever without being pushed. At every level of complexity, from molecular to

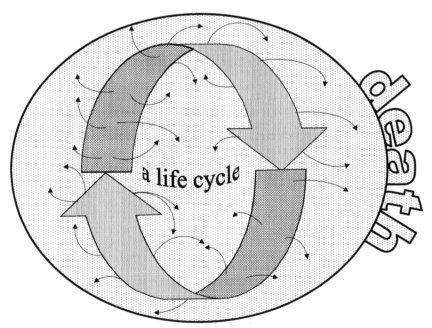

FIGURE 3.1. A lifecycle, as proposed by Thomas Huxley.

ecological and from cell to organism, at many points and in many ways, from inborn errors and outside pressures, smothering excess and asphyxiating absence, pollution and predators, accidents and hazards friction threatens life's momentum. At the same time, the catabolism of resources obtained from the environment pushes lifecycles through bioenergetics, replication (DNA-dependent DNA synthesis), transcription (DNA-dependent RNA synthesis) and translation (RNA-dependent protein synthesis), development and differentiation, growth and maintenance, anatomy and physiology, behavior and reproduction.

Most of what one thinks of as life's adaptations occurs on the power side of lifecycles, namely, the resources that give life more bounce to the ounce. Life's devices for reducing friction, on the other hand, are less obvious. Thus, living things constantly dissipate energy and excess, waste and wreckage (the thin arrows in figure 3.1), all of which would otherwise drag down lifecycles and threaten life's momentum. The environment absorbs many potential sources of drag, frequently recycling them back to life, but death does the lion's share of removing drag. Death reduces life's detritus to corpses that can be shed from lifecycles with a minimum of loss, disruption, and injury. In other words, *death is the milieu that lubricates lifecycles.*

TYPES OF LIFECYCLES

Lifecycles are *not* like other cycles in one way: lifecycles consist of two or more parts that must connect (hence the two open arrows in the figures 3.1 and 3.2 rather than an unbroken circle or one continuous arrow). The two parts may represent activities such as replication and transcription, one cell dividing into two (binary fission), or parents producing offspring, but there will always be at least two parts. What is more, organisms, whether single- or multicellular, going through a part of their lifecycle go through different periods or stages during that part (figure 3.2), and these organisms, while going through these periods or stages, are vulnerable to death in different ways, especially while transiting between parts, periods, or stages.

Lifecycles, like all other cycles, are without beginnings or endings. Indeed, life as we know it has not had a beginning for nearly four billion years,[2] and one can only guess what got life going then.[3] One cannot even say that a lifecycle ends with the extinction of a species, since evolution would seem to perpetuate lifecycles even then.

Of course, one should not assume that all lifecycles are the same (see appendix) any more than one should assume that all life is the same. Species go through different lifecycles as well as having different appearances. Life may have had several beginnings[4] despite the evidence for life's singularity, namely the near universality of the genetic code. All organisms utilize virtually the same three base codons in messenger ribonucleic acid (mRNA) for encoding the same amino acids in proteins and much the same start and termination codons as well.[5] But the near-universal code could have spread to originally different life forms. Evolutionary theory does not exclude the possibility of one code devouring other codes or displacing them (known as "introgression"). Indeed, reticulate evolution provides for mixing of genes introduced by recurrent hybridization.[6]

Similarly, different lifecycles may have sprung from different originary lifecycles rather than emerging as variations on a theme. Moreover, death in one lifecycle may differ from death in another lifecycle as much as life itself differs. Similarities in extant lifecycles may represent conserved qualities (biological inertia) or convergences among originally different styles of lifecycles. Differences may represent variations on an original theme or the conservation of original differences.

In general, three types of lifecycles are easily distinguished: lifecycles of noncellular life (viruses, bacteriophage), noncompartmentalized cellular life (prokaryotes: Archaea and Bacteria),[7] and compartmentalized or nucleated cellular life (eukaryotes).[8] Briefly, the lifecycle of noncellular life consists of two frequently overlapping parts: a diffuse, obligate parasitic part and a particulate or infectious part that may be "free-living" or reside in a cell. Classically, an infectious viral particle moves from one host cell to another, while the parasitic part reproduces and leaves the host as an infectious particle. The lifecycle is complete when the new infectious particle parasitizes a new cell, but infection can occur passively when already-parasitized cells, having already replicated their own genes and representatives of infectious particles, divide into new cells.

The lifecycle of cellular but noncompartmentalized life forms also consists of two frequently overlapping parts, namely, growing cells and dividing cells. Although the rate at which a prokaryote moves through these parts can vary greatly, the cell does not ordinarily stand still. Cells that fail to move through both parts of the lifecycle continuously die and fall into death's sink. The exception is dormancy; for example, through endospore formation, in which case a cell may remain metabolically inert, it would seem, virtually indefinitely.[9]

Some biologists have suggested that viruses, and even prokaryotic and small eukaryotic cells (such as algae, protozoa, and yeast) are immortal because they seem to divide indefinitely. But these biologists have confused reproduction and cycling: reproduction is making new organisms, while life cycling is rotation through life's parts or phases (for example, the growth and budding of yeast) and stages (e.g., the embryo, fetus, newborn, and so on, of placental mammals). Viral reproduction, of course, depends on parasitizing cells and cannot, therefore, be considered a form of immortality any more than the parasitized cells are immortal. Reproduction of prokaryotes and small eukaryotes is almost inevitably a pathway to death inasmuch as division is always asymmetric in one way or another, and one cell will inherit the more vulnerable, older part of the original cell, the part with "cumulatively slowed growth, less offspring biomass production, and an increased probability of death,"[10] while the other cell will inherit the new, more viable, and fecund part. Even the apparently symmetrical divisions of cells of the bacterium *E. coli* are not entirely symmetrical.[11]

Thus, cells as such are not the same after dividing. Some will age and undergo senescence, and cannot be said to be immortal. Indeed, the different products of cell division may even accumulate at the same time and place, much like the mortal, somatic cells of multicellular eukaryotes. Moreover, even the heartiest viruses and cells are hardly immortal when they stand condemned on death's doorstep due to failures to complete the growing part of a lifecycle.[12]

THE EUKARYOTIC SEXUAL LIFECYCLE

The cellular and compartmentalized forms of life known as eukaryotes (truly nucleated)—especially their multicellular representatives, including us—have generally added sex to their lifecycle, although some eukaryotes have either lagged behind and lack sex or have lost sex from their repertoire of life's pursuits and merely cycle through growth and division.[13] In eukaryotes with a sexual lifecycle, every spin of the cycle makes a new generation as well as a new organism.

Where sex enters the cycle, one part of the lifecycle is identified with germ or sex cells (egg and spermatozoon) and the other part with somatic or body cells and stages of a life span. The somatic part is broken into periods—prior to, during, and, frequently, after sexual maturation—and most sexual lifecycles take on additional complexity identified by stages within periods. The transitions between each of the parts, periods, and stages that comprise sexual lifecycles are also nodes in the wave towards death. Parts center on the production of germ or sex cells (also known as gametes) via oogenesis and spermatogenesis, while stages commence with fertilization, from fertilized egg through adult. In both parts, cells or organisms seem to have especial difficulty moving through particular nodes and stages (for example, the gastrula stage).

In our case, a pre-maturation period commences when an egg and spermatozoon fuse and ultimately, but not immediately, form an embryo (that is, the preembryo or blastocyst contains no embryo until implantation is virtually complete), followed by a fetus, a neonate or baby, and a juvenile. At none of

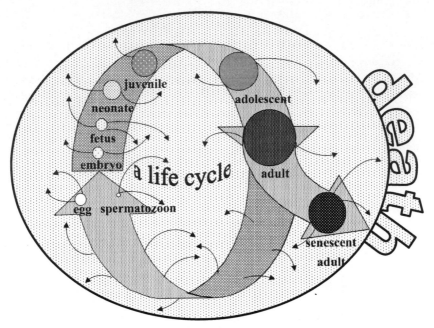

FIGURE 3.2. Death's role in a life cycle.

these stages are we fertile. The reproductive system matures at puberty, corresponding to the cessation of bone growth. Adolescence "is part physiological, part psychological, part social construct"[14] and may extend to 19.5 years in women and 20.9 years in men if sleeping patterns are taken into account.[15]

Maturity and sexual prowess reach their pinnacle in the adult stage, when the differentiated products of the germ line, egg and spermatozoa, have the greatest probability of successful fertilization and continuing the lifecycle. Finally, older adults become senescent adults and may enter a sterile, postreproductive period (in the postmenopausal period in women) or a period of diminished sexuality (age-related impotence in men).

In terms of death, the early pre-maturation stages are especially vulnerable; the juvenile is especially hearty; while death rates become exponential following adolescence until they level off in senescent adults. In addition to all the other causes of death threatening adults, senescent adults also suffer from so-called natural death. Senescent adults exhibit several adaptations that would tend to prolonged life, such as reduced metabolic rate and appetite as well as whatever immunity may be derived from having navigated through life successfully up to that point.

SOME DETAILS OF THE EUKARYOTIC LIFECYCLE

Many qualities distinguish eukaryotic cellular life from prokaryotic cellular life, but, of all these qualities, the most conspicuous is the high concentration of hereditary material or DNA within the compartment known as the nucleus and the isolation via the nuclear envelope of this DNA from the so-called cytoplasm. What is more, unlike prokaryotes, in which DNA replication is continuous and cells divide as their load of DNA is replicated, in eukaryotes, DNA replication is discontinuous, confined to a phase known as the synthetic or S phase, and cells cycle through so-called gaps or G phases, G_1 after division and before S, and G_2 after S and before divisions. Thus, eukaryotic cells carry different loads of nuclear DNA, one load after division and before S, and two loads after S and before division.

Moreover, the nuclear DNA of eukaryotes is wrapped around nuggets of basic proteins (histones) in nucleosomes and joined by additional nucleoproteins in complexes called *chromatin* that condense prior to cell division into *chromosomes*. Eukaryotic chromosomes are duplicated following the replication of DNA and separate into identical *chromatids* through a process called *mitosis*[16] (also called *karyokinesis* or nuclear division), resulting in the formation of nuclei with matching sets of chromatids. Chromosomal condensation and mitosis are uniquely part of eukaryotic cell division. The division of the cytoplasm (called *cytokinesis*) may follow mitosis completing division of the eukaryotic cell. Alternatively, cytoplasmic division may be delayed, allowing nuclei to accumulate within a unified cytoplasm (known as a *plasmodium*), or both nuclear and cytoplasmic division may be suppressed following the fusion of cells into a common mass (known as a *syncytium*).

The separation of replication from cell division by gaps seems to have created opportunities for the eukaryotic cell to differentiate and to introduce sex into its lifecycle. Differentiation and sex are also unique to eukaryotes. Ordinarily, eukaryotic cells differentiate in the gap (G_1) after cell division and before the replication of DNA, while cells undergo sex in the gap (G_2) between replication and cell division. *Differentiation* affects the cytoplasm of eukaryotic cells, while *sex* is effectively a nuclear process.

Breaking the eukaryotic lifecycle open and spreading it out as a wave over a time axis helps one visualize the sexual lifecycle. Eukaryotic cells with a single dose of hereditary material (represented by thin arrows in figure 3.3) move through half of the wave, and cells with a double dose of hereditary material (heavy arrows) move through the other half. Cells with the double dose are said to be *diploid*, while those with the single dose are said to be *haploid*. In *fertilization* (also known as conjugation and coupling), two haploid cells fuse and give rise ultimately to diploid cells, while in *meiosis* (diminu-

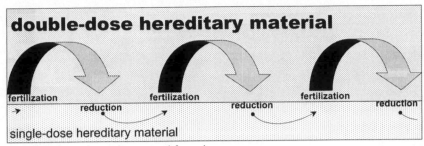

FIGURE 3.3. The Eukaryotic lifecycle as waves.

tion) or *meiotic reduction*, a diploid nucleus divide twice to give rise to haploid nuclei with half the original amount of hereditary material.[17]

In the broad sweep of eukaryotic organisms, cells may divide in both parts of the lifecycle, forming a clone, plasmodium, or multicellular body (*soma* or *thallus*), and cells may differentiate in both parts, forming different types of cells. In our case, as in the case of all multicellular animals, the cells in both parts of the lifecycle differentiate, but only cells in the diploid part divide and form a soma. The haploid cells only differentiate as sex cells known as eggs (or oocytes) and spermatozoa (singular: spermatozoon). Each individual human being is (largely) a soma formed by an aggregate of differentiated diploid cells, plasmodia, and syncytia that, through their collective activities, perform all the functions equated with living.

Because of the great disparity in the haploid and diploid cells of multicellular animals (single versus multicellular; lacking cell division versus having wave upon wave of cell division; differentiation into one or two cell types versus differentiation into more than two hundred cell types; etc.), haploid cells are often thought of as merely linking diploid parts of our life wave through sexual reproduction. For us, haploid cells are thus the *germ* (cells that may develop) inasmuch as these cells participate in sexual reproduction and the creation of diploid organisms. But these same haploid cells only arise via meiotic reduction and differentiation of originally diploid cells that should, therefore, also be considered the germ.

In other words, a germ, if it is definable, consists of both haploid and diploid cells rather than purely one or the other. Likewise, the body of a reproductive adult animal, with few exceptions (cnidarians, tunicates, and flatworms), contains a preponderance of diploid cells but also some haploid cells. The two parts of the life wave, therefore, are only loosely equated with germ and soma.

In the seesaw of eukaryotic life waves, the amount of hereditary material passing through haploid and diploid phases is held in balance by coupling fertilization with its opposite, meiotic reduction. Fertilization, the fusion of two haploid cells, is followed by a complex process of chromosomal mingling and cell division, leading to the formation of diploid cells. Meiotic reduction, on the other hand, is the complex process involving one round of DNA replication and two cell divisions leading to the formation of (as many as) four haploid cells. These haploid cells differentiate into spermatozoa in male animals and pollen in male parts of plants, while one or more ova form from the products of meiosis in female animals and ovules in the female parts of plants.[18] In algae, ferns, liverworts, and mosses, the immediate products of meiotic reduction are spores that, following their development into multicellular organisms called *gametophytes*, eventually produce germ cells known as *gametes*. In protozoans and fungi, haploid germ cells are also called gametes and may differentiate from gametocytes (for example, the malarial parasite in its mosquito host). In ciliates, diploid cells may play the role of gametes and meiotic reduction as well as fertilization may be reduced to nuclear divisions and exchange.

What happens to cells in the two parts of the life wave—between fertilization and reduction and between reduction and fertilization—differs in different eukaryotes depending on whether the organism is unicellular or multicellular, and whether it is a plant, fungus, or animal. Evolution has molded both parts, and different groups of eukaryotes exploit one or the other part. In both parts, however, to the degree they are present, cell division, cell migration, cell death, and the synthesis of various cell products determine the structure and function, morphology and physiology, development and maintenance of a unicellular colony or multicellular thallus or soma.

The major difference among living eukaryotes is determined by the degree to which each lifecycle exploits its haploid and/or diploid parts. Microscopic algae, fungi, and many protozoa reserve the right of prolonged cell division and differentiation for one, the other, or both parts of the lifecycle, but in macroscopic eukaryotes (with the exception of mosses and liverworts), life's conspicuous roles are performed by the diploid soma. Large multicellular organisms, such as human beings, exploit the part of the lifecycle consisting of diploid cells while suppressing the part of the lifecycle consisting of haploid cells.

We consist almost entirely of diploid cells and syncytia (skeletal muscle, giant cells of the placenta), with rare plasmodia (aging liver and urinary bladder epithelia) and exceptional haploid cells. These haploid cells are relegated to our gonads (ovary or testis) and differentiate into eggs and spermatozoa without cell division. Moreover, these cells are doomed to a short existence (on the order of hours to days, although in bats and bees they last much longer). The haploid cells of animals, thus, constitute a self-limiting cell line whose existence in every generation depends on regeneration from diploid cells following meiotic reduction. Similarly, trees consist almost entirely of diploid cells, although as many as eight haploid cells (equivalent to a highly reduced gametophyte) may be produced in flowers by the division of haploid cells prior to fertilization. Indeed, a fundamental difference between animals and plants is the elimination of cell division in haploid cells of animals as opposed to the retention of some cell division by haploid cells of plants. In animals, moreover, the cells capable of reduction generally comprise a single line of cells called the germ line (with cnidarians, tunicates, and flatworms being exceptions), whereas in plants, virtually any cell capable of division may also be capable of undergoing reduction (there is no self-limiting germ line).

Of course, life continues (or cycles) only by passing through both parts of the life wave—the haploid and diploid parts—and, of course, death also occurs in both parts. Any cell, whether haploid or diploid, can die.

The greatest misunderstanding of biological life arises from thinking that life occurs only in the diploid part of the lifecycle and only organisms in that part are alive.[19] The reason for this misunderstanding may reflect a sort of large-organism chauvinism or a myopia in which large organisms alone are "seen." Many biologists have tried to rectify the consequent confusion with the help of the microscope. As early as the seventeenth century, biologists began to clarify the role of "animalicules" or "zoe," later known as germ cells, such as eggs and spermatozoa, and by the late nineteenth century, biologists understood the roles of meiotic reduction and fertilization in pasting together the parts of the lifecycle or wave.

Regrettably, many twentieth-century commentators on life have persisted in the notion that life begins at fertilization. In fact, fertilization is not a beginning of life, since the egg and sperm are also alive at the time they fuse to form a fertilized egg. Fertilization is important enough in its own right, but it should not be confused with the start of life for several reasons.

Above all, the processes occurring at fertilization are part of a continuum, not a rupture. First of all, in our case and that of most mammals, the *oocyte*, or maturing egg, has not even completed the nuclear divisions that reduce its amount of hereditary material by half (known as the second meiotic division) at fertilization. Second, only a *heterokaryon*, or cell with two different nuclei, called the *zygote* (meaning yoked) is formed following fertilization: the nucleus known as the *female pronucleus* is formed with half of the egg's hereditary material and the *male pronucleus* is formed with the sperm's DNA but with other ingredients derived from the egg. Third, and most importantly, the female and male pronuclei do *not* fuse (with rare exceptions such as the sea urchin). Rather, DNA in the two pronuclei undergoes replication, and, subsequently, as chromosomes condense, the nuclear envelopes (membranes) interdigitate and break down, and naked chromosomes mingle on the division (metaphase) plate of the zygote's cleavage (division) plane. Our zygote never has a single, unified nucleus (chromatin within a nuclear envelope). Instead, the first nucleated cells in the new diploid part of the lifecycle are *blastomeres* formed by the division of the zygote.

Other arguments also require abandoning any idea one might have of life forming at fertilization. For example, a fertilized egg is sometimes a "blighted ovum," giving rise to a preembryo or blastocyst (and even persisting as a chorionic sac for several weeks) but with *no embryo* within it at all! Furthermore, identical (monozygotic) twins arise in about thirty-six of every ten thousand births. Thus, at least on rare occasions, two individuals, not one, would arise from a single fertilized egg!

STAGES OF A LIFETIME

Rather than inundating the fertilized egg with properties it does not have—such as the beginning of life—one might consider fertilization the prelude to the diploid phase of the eukaryotic life span and to development in the eukaryotic lifecycle. The zygote is followed by the preembryo or blastocyst and that by embryo and fetus, leading to neonate and all the subsequent stages of the diploid lifetime. Stages are not pure conveniences, human inventions for stratifying the diploid part of a lifecycle. Rather, stages are well-characterized episodes that take place in particular places—environments—with specific durations, attributes, and, most importantly, vulnerabilities or probabilities of failure and death.

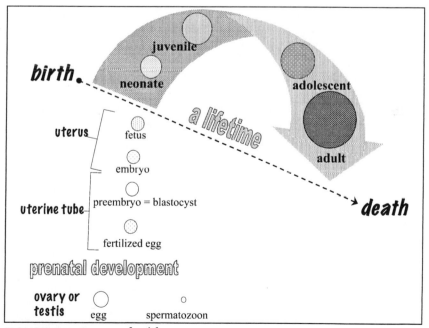

FIGURE 3.4. Stages of a life span.

In the case of mammals, including us, eggs develop in ovarian follicles within an ovary; spermatozoa develop in seminiferous tubules within a testis. Of the forty or so eggs developing each month of the human ovarian cycle, typically, all but one will die before reaching maturity. Likewise, male germ cells die in the course of their development and after ejaculation.

Preembryo

Fertilization takes place and the fertilized egg is formed and resides within the uterine tube, where, covered by a "shell" or *zona pellucida*, the zygote and subsequent blastomeres undergo cell divisions (known as *cleavage*) and form a multicellular *blastocyst* or *preembryo* consisting of a *trophectoderm* containing an *inner cell mass* (ICM). After entering the uterus, the blastocyst spends several days in suspension before hatching, attaching, and implanting. The blastocyst proceeds to form villi on its surface while solidifying the pregnancy and inducing maternal cooperation in the formation of the placenta.

Of the four hundred or so eggs a woman will bring to maturity in her lifetime, the overwhelmingly vast majority fails to be fertilized. The probability that any fertilized egg will develop is estimated to be as low as 20 to 30 percent[20] with large numbers of those developing to blastocysts failing at the point of implantation and many of those successfully implanting failing to gastrulate and form an embryo. The woman involved merely goes on to menstruate, possibly having a heavier period than usual.

If implantation is successful, however, the blastocyst, while still lacking an embryo as such, rushes to divert vast numbers of its cells to villus formation, thereby establishing sources of nutrition and sites of gas exchange in the uterus. Now implanted, the blastocyst produces hormones of pregnancy (human chorionic gonadotropin) and, at the same time, continues to provoke the formation of the maternal part of the placenta. Only then, a month after the last menstrual period and fifteen days after fertilization, is the first suggestion of an embryo's development (or two embryos in the case of identical twins) apparent in the *epiblast*, a thin cell layer running across the blastocyst's cavity. There, a transient thickening known as the *primitive streak* puckers upward; a primitive groove forms within the streak and identifies the future embryonic axis while marking nothing more than the vicinity of the future anus. Between sixteen and seventeen days after fertilization, a genuine bit of the embryo takes shape ahead of the primitive streak: superficially a translucent notochordal process appears which will subsequently thicken and round up into the definitive notochord.

Embryo

Internally, cells have moved en masse through the primitive groove and into extensive extraembryonic membranes and a minute embryo. The segregation of these moving cells through the process known as *gastrulation* accounts for the production of three *germ layers*: ectoderm, endoderm, and mesoderm. Gastrulation proceeds to the neural plate stage at eighteen to nineteen days after fertilization. Then, the neural plate thickens rostrally and the primitive streak, having exhausted its font of cells, regresses caudally.

Gastrulation is, undoubtedly, one of the most difficult processes the organism will ever have to negotiate, and many gastrulas fall by the wayside. As Lewis Wolpert, the embryologist and media personality, has indicated, "It is not birth, marriage or death, but gastrulation which is truly the most important time in your life."[21]

The rates of early pregnancy loss (EPL) or death at early stages of development are difficult to estimate but frequently notoriously high.[22] Some data suggest that "[o]nce pregnancy is recognized clinically, it is accepted that 12% to 15% undergo spontaneous abortion. . . . [But only about 3% of women with] viable pregnancies at eight weeks . . . experience fetal loss thereafter."[23] Other data indicate that the "maximum chance of a clinically recognized pregnancy occurring in a given menstrual cycle is 30–40%."[24]

Fetus

At the beginning of the ninth week of pregnancy, or nearly two and a half months after the last menstrual period, a fetus replaces the embryo, complete with rudimentary organ systems. These systems function in their own way, supporting the life and development of the fetus but hardly functioning the way their namesakes will function following gestation. But slowly, if not surely, development proceeds until, at birth, a neonate emerges—not quite ready to take on the world, but demonstrably capable of living in the terrestrial environment if only with considerable assistance.

Recently, as a consequence of the widespread use of sonography for the diagnosis of pregnancy, suspicion has arisen that as few as one out of ten to twelve pregnancies involving dizygotic (fraternal) twins results in twin births. One of the two twins are thought to be absorbed, occasionally by the surviving twin, which results in a chimera having cells, if not organs, derived from the absorbed twin.[25] Other evidence suggests, however, that the "presence of one embryo . . . does not impede the development of its twin"[26] when two eggs are fertilized and develop at the same time.

Neonate

Following birth, it might seem that the neonate has suddenly come alive, but the neonate also remains quite vulnerable to death. The neonatal stage—roughly the first year—is frequently divided into two periods, an early neonatal period of the first twenty-eight days with a higher death rate than the post-neonatal period from twenty-nine days to one year. As late as 2001, statistics for the United States showed that for every 1000 live births, 4.5 early neonates and 2.3 post-neonates died.[27]

Post-neonate, Toddler, and Juvenile

The post-neonate, or infant after the first 28 days, is less fragile than the early neonate, and infant fragility is further reduced during the first year and subsequently as the toddler achieves mobility. The juvenile, or preadolescent, is still more independent—and demanding—growing and acquiring skills at mind-boggling rates, but, most importantly, *fragility declines to the lowest level in the lifetime.*

Adolescent to Senescent

And then comes adolescence and sex, along with physical prowess, colossal self-confidence, and titanic self-esteem. Numerous misunderstandings experienced in families can be attributed to differences in social adaptations of toddlers, juveniles, adolescents, and adults, each occupying their different niches. Life is continuous but not homogeneous, and we are not the same individuals as we pass through life's stages, exchanging environments and morphologies.

Beginning with adolescence, the lifecycle may be completed through the differentiation of eggs or spermatozoa capable of undergoing productive fertilization, although, in Western culture, child bearing and rearing is typically delayed until a more full-blown stage of adulthood. Maturity sets in, and the young adult begins to get a fuller picture of life, while the older adult becomes weighed down by life. Throughout the period between adolescent and senescent adult, however, one thing is constant: life expectancy declines at an exponential rate. Indeed, the mortality rate doubles every seven to eight years, until senescence when the rate levels off at something in the vicinity of 50 percent per year.

Adulthood is a period of decline, but irreversible decline actually began long before the full-grown adult stage. Decline begins at adolescence before anyone notices, and continues throughout adulthood, although we may take extraordinary measures to overcome and deny the decline. Of course disease and accident can strike one down at any stage along the way, and many of us will not live long enough to experience senescence, but it is there for those who survive all the other stages of a lifetime. During senescence, decline becomes unmistakable, and through senescence the body moves toward "natural" death.

ADAPTATIONS TO LIFECYCLES

From the human vantage point, a lifetime is defined by the interval between birth and death. Development and maturation occur early in the lifetime, while aging occurs later, and dying occurs at the end, whenever that may be. Darwinian natural selection would predict that everything about such a lifetime would be adaptive, and one hardly has any difficulty imaging how development and maturation meet the requirements for adaptation. The question is how aging, dying, and death can fit into the vice of adaptation?

AGING

Definitions of aging quickly become mired in stagnation and, hence, lose contact with the vitality of life. Aging, nevertheless, is a quality of life even when misconstrued. For example, *no* significant loss of hippocampal or neocortical neurons accompany aging (as opposed to interneuronal signaling), and old dogs *can* learn new tricks.

The fact that aging works adaptively on behalf of life extension would be obvious were it not that aging is first thought of as a symptom of approaching death. For example, while declining metabolic rate and core body temperature are considered age-related degenerative changes of physiological functions, the same changes induced by caloric restriction in young individuals are correlated with prolonging life and good health. Ironically, the decline in appetite that accompanies aging is thought of as death's harbinger instead of an instinctual promotion of caloric restriction. Moreover, age-related changes in the endocrine system, such as those affecting the reproductive system, may be a blessing in terms of Darwinian fitness rather than a curse in terms of weakened physique. Specifically, the postreproductive period in women (post-menopause) can sometimes be used to promote the reproductive success of offspring and the survival of grandchildren (the so-called grandmother effect. See chapter 6).[28]

On the other hand, nonadaptive changes typically ascribed to aging may be more nearly related to other factors. The "metabolic syndrome," including disturbed fibrinolysis, hypertension, dyslipidaemia (increased serum triglycerides) and impaired glucose tolerance/type 2 diabetes mellitus (increased insulin resistance), and reduced bone mineral density may be more nearly related to obesity, poor diet and habits of exercise, than age.[29] So-called age-related disorders in cardiac function, including heart rate, blood pressure and arterial stiffness, are certainly not correlated with age alone. Diet, especially high-fat diets, and smoking may be the prime culprits in some chronic diseases, especially atherosclerosis and other cardiac pathologies: aortic valve calcification, congestive heart disease, hypertrophy of cardiac muscle, and interstitial fibrosis. High-fat and high-caloric diets may also be correlated with diabetes. Some of the gravest structural damage attributed to aging may be more closely correlated with "wear and tear": diminished stature, decline in bone mineral density at particular sites; and osteoarthritis, especially in fibro-cartilage between vertebrae and the articular cartilages of synovial joints.

Which is not to say that aging is perfectly innocent. Aging is a major player when it comes to vulnerability or frailty. For example, the rate of wound healing and the resilience of skin both decline with age, and declining func-tion in the immune system may be responsible for increased susceptibility to infectious and autoimmune diseases and cancer. What is more, the metabolic syndrome may be exacerbated by age if fat mass, especially abdominal fat, increases, while lean body mass declines.

But aging as such may not be the leading cause in the etiology of behav-ioral deficits associated with changes in the brain: hippocampal dysfunction, decreased cerebral blood volume, lose of dendrites and arbors in the cerebral cortex, diminished affinity of neurotransmitter receptors (for example, dopamine receptors) and transporter binding proteins (presynaptic vesicular acetylcholine transporters), loss of hippocampal cholinergic fibers, and alter-ations in white matter. The similarities between the pathologies of these so-called age-related dementias and various forms of mammalian spongiform encephalopathies as well as Alzheimer's disease (amyloid β plaques and neu-rofibrillary tangles) suggest that age-related behavioral deficits in general may be more a function of disease than of aging as such. Indeed, the aging indi-vidual may have been waging a heroic battle against the encroachment of dementias for a long time before symptoms erupted.

Dying: The Adaptation

Typically, one says individuals are dying when they are at death's door, and dying is considered a prelude to death and equated to morbidity. But dying occurs in living things throughout a lifetime. Indeed, the problem with con-temporary conceptions of dying is that they are dominated by notions of inert-ness and stasis associated with death rather than with action and dynamics associated with life. Even at the cellular level, dying may be "a protective mechanism" on behalf of the organism.[30]

Dying is a process. It is the systematic walling up and shutting down of injured and malfunctioning systems. Dying is, thus, an adaptation—a last-ditch adaptation, to be sure, but one that works, at least on occasion—to local-ize injury, confine damage, prevent the spread of toxic substances, halt the advance of death, and give the individual a chance to heal and recover.[31] When successful, the "dying" individual may not "die" at all! Indeed, dying individuals who manage to heal and regenerate are said to have "survived," as if they hadn't been dying in the first place. What should be said of these individuals is that the adaptive part of dying kicked in on time, preserved life, and postponed death.

In common usage, "dying" is so remote from its biological function that other words are used to signify dying's adaptive qualities. Words such as "crisis" and "conversion"—the turning point in the course of a disease when the patient will either improve or sicken further—connote the confrontation with survival. "Amphiboly" might also be considered as a substitute word for "dying" to convey the sense of a body's going in either of two directions or having qualities of both life and death.

Even when death is the climax of dying, death does not fall simultaneously on all the body's parts. Death is only the last stage in the catastrophic shutting down of body parts and ultimately of the entire organism—the "limiting stage in the organism's transition from the normal state to the 'one foot in the grave' state."[32] Indeed, the staggering of bodily shutdown provides the rationale for rescuing transplantable parts of a dying body.

Somewhere between localized, reversible shutdown and complete, irreversible shutdown lies a state of incomplete shutdown. Therein, the life of an individual is no longer tenable, but parts of the individual might continue functioning were they transplanted to another being. Managing to keep these parts alive prior to transplantation has been a miracle of medical technology, and making death compatible with transplantation is a masterpiece of juridical legerdemain. There are many stages of death, as the anthropologist of death, Phillipe Ariès, among others, explains: "there is brain death, biological death, and cellular death. The old signs, such as cessation of heartbeat or respiration, are no longer sufficient. They have been replaced by the measurement of cerebral activity, the electroencephalogram."[33]

Many a transplant recipient is grateful to have received an organ, and, no doubt, many more candidates would be happy to receive organs, but few adults carry a donor's card (permission to remove and transplant organs following the cessation of cerebral activity). A massive change in public attitudes toward death would have to take place before carrying a donor's card became customary. Above all, life would have truly to be considered a gift to be dispersed wherever possible and not a possession to be hoarded as long as possible.

DEATH: THE ADAPTATION

Death, or the termination of a life span, occurs only in living things. Only organisms and cells with their contents intact can die. Chemical reactions do not die; they only reach thermodynamic equilibrium or an end point resulting from the loss of reactants through a phase shift. This intimate relationship of death to life makes sense because death is the chief device available to life-cycles for reducing the friction that might slow life's momentum were unviable detritus to stick to life. Death functions in life by way of breaking off life's detritus.

Ironically, most of death taking place in and around living things is not readily seen or appreciated as death. Death is recognizable when seen at the level of organisms even if it is not accepted with quiet equanimity. For example, death is acknowledged when a Mack truck has flattened someone or the truck's equivalent—pathological microbes or arrant cells—has demolished someone from the inside out. Certificates of death declare that an individual has died from one or more recognized cause(s), a notifiable disease and its sequela or some generalized clinical diagnosis.[34] Most frequently, human death is attributable to disease affecting the gut and associated organs, the blood or blood-forming organs, the circulatory systems, and, increasingly less frequently, the endocrine, urinary (kidneys), muscular and skeletal, nervous, or reproductive systems.[35] Remarkably, "old age" is not entered on death certificates as a cause of death, although widely recognized by symptoms. Instead, pneumonia and congestive heart failure, among other proxies for old age, are the pronounced cause of death.

But death is not seen in its microscopic appearance. Death is not recognized at the cellular level. Death is the fate of most differentiated cells. Hair, for example, is made of dead cells (differentiated keratinocytes) even if one does not ordinarily think of hair as dead. One is more likely to think of cellular death when seeing dandruff—the desquamated products of the dead inner root sheath of hair follicles plus the dead, secreted cells of sebaceous glands. Actually, all our body surfaces—both inner and outer surfaces—are constantly shedding dead cells, while internally, with the help of spleen and liver, our body sheds vast numbers of blood cells[36] and lymphocytes. Normally, fresh cells replace the dead cells and life depends on this turnover of cells, but life would simply be unsupportable were all the unviable cells produced everyday not dispatched as corpses.

We recognize death when it is rapid, as in the case of massive trauma or acute disease, when systems crash like computers or power grids. And we acknowledge death when it is slow, in the case of compact trauma or chronic disease and aging. Initially, some vital function is damaged or retarded, provoking the shutdown of local tissue. But like dying, we tend to ignore death if the individual at death's door is able to isolate the threat and give the body a chance to mobilize its repair or regenerative powers, which are prodigious, thereby restoring the body to health. If damage spreads, cascades, or avalanches, however, encroaching on additional functions, life may be incapable of compensating for lost functions or accommodating, and death prevails. We might then even blame the individual for dying! ("Well, he smoked all his life!")

But death at the cellular level is not understood as vital to life. Death at the cellular level does not inspire questions such as, "Who the slayer?/Who the slain?" One hardly makes a moral judgment about dandruff, although one may make a pragmatic judgment and change one's shampoo. And, although an invasive-destructive tumor and a fulminating infection may be thought of as enemies and threats, one hardly thinks of cancer or bacteria as invaders from the evil empire. Death at the cellular level is simply closer to life, and unlikely to invoke principles or incite rage. Moreover, to be fair, a part of our pain and suffering during advanced disease—ulceration, swelling, and fever—is due to the operation of our own bodily armor marshaled against the attack. But, ultimately, disease at the organismic level, especially diseases of the gut, blood and lymph, as well as aging and dying, are consequences of the failure to replace lost cells with fresh, healthy ones. Death may occur at different levels of complexity, but they are not unrelated.

Throughout one's lifetime, thus, death has been the endgame for cells taking part in normal, physiological turnover. Ordinarily, in the fullness of life, the very cells that fed into that endgame, namely adult stem cells, (see chapter 5), have been exhausted and are no longer available. At the same time, the reserve cells that might have replaced cells lost through trauma or disease are also exhausted. Death, it would seem, is not so much caused by the loss of stem and reserve cells as an acknowledgment of those losses. Ideally, death is efficient and few stem and reserve cells are left behind in the corpse.

THE LINEAR VIEW OF LIFE: LIFE'S ARROW

Raymond Pearl (1879–1940) ended his groundbreaking book, *The Biology of Death,* with a chapter on "natural death" in which he makes the astonishing declaration that "natural death is a relatively new thing, which appeared first in evolution when [a soma made by the] differentiation of cells for particular functions came into existence. Unicellular animals are, and always have been, immortal. The [germ] cells of higher organisms, set apart for reproduction in the course of differentiation during evolution, are [also] immortal."[37] Only the rest of the organism, the soma, is mortal. Like a good soldier, the soma serves, and is ready to lay down its life at the command of its master, the germ. Above all, the soma battles against the environment in order to protect germ cells and propel them into the next generation through sexual reproduction.[38]

Pearl's remarks evoke an ironic wince. In the period following World War I, Pearl was highly critical of "German Intellectuals" for arguing on behalf of "*Allmacht,*" or violent competition in the survival of the fittest, and hence of German nationalism and war. Pearl did not acknowledge, therefore, his blatant borrowing of ideas on death and evolution from the German zoologists August Weismann (1834–1914) and Ernst Haeckel (1834–1919).

Weismann, who was Professor of Zoology at Frieburg, had begun theorizing on lifecycles as early as 1882 when he claimed, in *Ueber die Dauer des Lebens*, that some strains of unicellular protozoans lived immortally by growing and dividing endlessly.[39] Weismann expanded this idea into the notion that eggs and spermatozoa, comprising the germ of multicellular animals, represent an immortal, unicellular part in the lifecycle of the otherwise mortal multicellular organism.[40] Weismann thus divided the lifecycle of multicellular animals between a single-celled germ that remained perpetually in a primitive state, and a multicellular soma doomed to mortality but evolving into a highly advanced, complex, and differentiated organism.

Some question may remain about the soma's effects on the germ during its sojourn in the midst of the soma, since aging in parents is not entirely neutral regarding the well-being and longevity of offspring.[41] Nevertheless, many modern commentators contend that the mortal soma we value so highly— namely us—"acts merely as a host for the perpetuation of immortal germinal material."[42] The soma's cells, tissues, organs, and organ systems may have evolved furthest from the primitive germ; they may give us form and substance as they differentiate and give us health as they resist the wear, tear, and strain of functioning, but the soma is also mortal and gives us death. Mortality, it would seem, is the tradeoff for specialization, activity, and robustness during a lifetime, but once the germ line has moved to the next generation, the soma left behind is dispensable.

Ernst Haeckel, who was known as the "German Darwin" and was Professor of Zoology at Jena, wrote the edict that legitimized the notion of the primitive and immortal germ versus the highly evolved but dispensable soma. Haeckel preached a doctrine of biological recapitulation that proved to be a formidable rationale for evolution in the epoch of World War I. This doctrine,

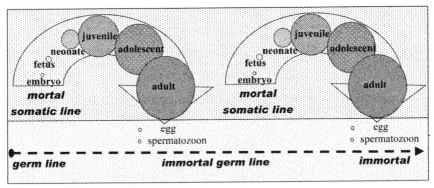

FIGURE 3.5. Weismann's divide.

that "ontogeny epitomizes (or recapitulates) phylogeny," decreed that an organism's development (ontogeny) condenses and repeats (recapitulates) its evolution (phylogeny), adding new evolutionary stages at the end of embryonic development while allowing a moderate amount of adjustment to accommodate new conditions. Haeckel enthroned the egg and sperm as the immortal germ line and the fertilized egg and zygote as the first cell in the organism's mortal somatic line. The organism, thus, returned, as it must, to its single-celled origins in order to launch its development in the path of its evolutionary history. Likewise, the ancestral gastrea, the source of all further animal evolution, is represented by the gastrula, the source of all further animal development.

Weismann and Haeckel were towering figures in German biology and both seemed destined for immortality. Indeed, today, biology's vast edifice stands on the foundation of Weismann's divided lifecycle and Haeckel's doctrine of recapitulation, especially with the gastrea and gastrula as the central characters in animal evolution and development. Together, Weismann and Haeckel wed immortal eggs and spermatozoa into primitive, single cells with an evolutionary memory able to dictate the development of an advanced multicellular organism. In addition, Weismann and Haeckel divorced the single celled germ line from the multicellular soma. While giving the soma responsibility for passing the germ from generation to generation, the soma became mortal and was allowed to die after performing its role in life.

MOLECULAR CONSEQUENCES

Weismann's place in mainstream biology was assured when the "germ" he popularized became synonymous with the self-reproducing nuclear determinants of heredity otherwise known as *germplasm*. Indeed, Weismann's version of "hard" heredity through germplasm is said by some (revisionist?) historians to have anticipated the highly touted "rediscovery" of Mendelian particulate inheritance and hence of the gene and DNA that followed.[43] Moreover, Weismann's concept of determinants passing from the nucleus to the cytoplasm is portrayed as a harbinger of mRNA, and Weismann is sometimes credited with having preceded Francis Crick in laying out the serial hypothesis (DNA → RNA → protein) connecting nuclear genes to cellular differentiation.

Haeckel, on the other hand, was only temporarily installed in biology's pantheon. His gilding tarnished in the wake of World War I and several attacks on the metaphysics of his doctrine of recapitulation.[44] But he lives on, if surreptitiously, in today's molecular version of recapitulation: the "blueprint," "roadmap," "*Bauplan,*" or rules for development supposedly written in the language of genes and inscribed in sequences of nitrogenous bases in DNA. The sheer chemical simplicity of DNA, coupled to elaborate editing mechanisms, guarantees the fidelity of the genes' reproduction, otherwise known as DNA replication, while small changes in DNA, or mutations, may occasionally prod the soma to develop slightly differently and thereby acquire some advantage for posterity.

According to contemporary biology's canon, the divide between the soma and germ is a divide between vegetative and reproductive functions. The divide does not depend on the mutual interactions between soma and germ, although there are many interactions. Rather, the divide allows the soma to evolve multicellular complexity while maintaining the germ in primitive, unicellular simplicity. This division of labor between the soma and germ permits the soma to attain extraordinarily large size—containing differentiated cells of numerous types, tissues, organs, and organ systems—adapted to nurturing the dormant germ and facilitating its passage from generation to generation. The massive jobs of development or autopoiesis and maintenance or homeostasis are performed by the soma through activities of the *somatoplasm* or cytoplasm, through cell division that installs the cellular pedigrees known as *somatic lines*, and the production of extracellular materials of all descriptions (from soluble hormones to the hard parts of bone and teeth). The sex cells, on the other hand, merely perform the function they have presumably performed successfully for hundreds of millions of years, namely, reprogramming the genome, preparing for productive fertilization, installing (fusiogenic) sites in membranes capable of fusing, and activating mechanisms for the development of the soma.

The Central Dogma

In addition to nurturing and transporting the germ, Weismann argued that the soma buffers the germ against deleterious influences by absorbing damaging, irreversible changes that arise from interactions with the environment and from products of differentiated cells, tissues, organs, and organ systems. In doing so, the soma suffers and eventually dies, but it also protects the germ line from these same deleterious influences and prevents their transfer to the germ line. Protecting the germ line may seem like the icing on the cake of the germ/soma separation, but it is of central importance. Protected by a firewall of soma, the germ is immune to changes wrought on the soma in the course of a lifetime.

Weismann was singularly dedicated to the dogma that the germ is insulated from influences that come to bear on the soma. He was fiercely opposed to the French teleological slant on adaptation given to evolutionary theory by Jean-Baptiste Lamarck (1744–1829). An inveterate anti-Lamarckian, Weismann was determined to show that life did not adapt out of experience, desire or convenience. Indeed, "Weismann had sounded the death knell of . . . the [Lamarckian] inheritance of acquired characteristics: with great conviction he had argued that the sex cells have an existence independent of the rest of the body, and thus no new heritable changes can come from habit, use or disuse, and the like."[45]

In the twentieth century, Crick transformed Weismann's notion of germplasm-immunity into the central dogma of molecular biology, namely, that the serial flow of hereditary instructions from gene to protein does not flow backwards. Crick's central dogma led biology down the path traveled by natural sciences in Europe since the seventeenth century. This was a path generally sanctioned by Christian belief and medical practice that showed the way to control and improve human destiny (manipulating well-being and health). DNA was not only omnipotent in its dominion over the soma, but in its detachment from life's trials. Through genes, DNA rivaled God's inaccessibility coupled to transcendent power over the soma.

WHAT IS WRONG WITH THE WEISMANN/HAECKEL DOCTRINE

The modern version of the Weismann/Haeckel doctrine is so successful that one is liable to become lulled into acceptance. Therein lies the rub: the central dogma is liable to come back to haunt us. Most importantly, the Weismann/Haeckel doctrine is built on false premises: the immortality of the germ line and the mortality of the soma. In fact, the germ line as such is not immortal and the soma is not necessarily mortal. All life forms die when they fall off the lifecycle, and all life forms remain alive while passing through the lifecycle.

THE MYTH OF AN IMMORTAL GERM LINE

In theory, if germ cells "were to age the species would become extinct; they therefore retain a vicarious immortality by virtue of their passage from generation to generation."[46] Much as clones of protozoa are supposed to descend endlessly from a single, vegetative, or asexually reproducing cell, the germ line is supposed to continue by cell division unabated through eternity. But all is not well with this view of germ lines or protozoa.

For more than a century, "[t]he question of whether Protozoa can continue to multiply indefinitely by binary fission [cell division] without decline of health or vitality has been the object of intensive research."[47] The present consensus among cognoscenti seems to be that "Weismann was wrong, and isolate lines of protozoans are not immortal."[48]

The case is made most strongly for the "hairy" unicellular protozoa known as ciliates. In the absence of sex, most, if not all, naturally occurring clones (descendents produced by cell division of one originary cell) of ciliates are doomed.[49] Under optimal conditions, in laboratories, strains of the ciliates *Paramecium* and *Tetrahymena* exhibit vigorous cell division for a period, but then the rate of division slows, and, ultimately, division ceases entirely. The clone enters a stage of so-called vegetative senescence followed by the death of all cells.

On the other hand, clones of ciliates, even clones entering vegetative senescence, are revived by sex in the form of either cross-fertilization or self-fertilization. The ciliate form of cross-fertilizing sex begins when cells gather together ("swarm") in large numbers and pairs of competent cells (known as isogametes) "conjugate"—the ciliate form of copulation or sexual coupling—fusing at their oral surface or mouth ends. Each cell's nucleus (or micronucleus to be precise) undergoes a complex set of divisions (including, but not limited to, meiotic reduction) until only half the amount of hereditary material remains in nuclei.[50] Haploid nuclei are then exchanged between the fused partners. These nuclei pair and, having completed this ciliate version of fertilization, cells separate, and the rejuvenated "exconjugants" sustain cell division again with renewed vigor.[51]

The ciliate form of self-fertilization, known as *endomixis*, begins when the ciliate's nucleus goes through the same reduction division as a conjugant without conjugation, and ultimately two of the cell's own nuclei fuse and restore an active, vegetative nucleus to the cell. The process would seem to resemble "selfing" in hermaphrodites—snails, roundworms, plants—although the process takes place within a cell as opposed to among cells of the same organism. In any case, the cell whose hereditary material is reformed by endomixis is as competent at vegetative growth as an exconjugant.

Weismann's notion of an immortal germ line in multicellular organisms is also bizarre since germ cells are not self-renewing. Rather, vegetative cell division stops abruptly and irreversibly in an animal's germ line as soon as the germ cells begin preparation for germ-cell differentiation (when they begin meiosis). What is more, the vast majority of differentiating and differentiated germ cells die without having partaken of fertilization. Indeed, death among the germ cells proceeds on a scale comparable, if not superior, to that found in somatic cells.

The Myth of the Mortal Soma

The notion of a mortal soma, like that of an immortal germ line, is also embedded in biology's metaphysics, as opposed to empiricism. When one takes a close look at real living things, plants and animals, both invertebrates and vertebrates, one notices that some exhibit so-called negligible senescence and indefinite life spans.[52] Take, for example, the Earth's longest living clones, such as a box huckleberry more than 13,000 years old, the 11,000-year-old creosote known as "King clone" growing in the Mojave Desert, and the 10,000-year-old clone of Rocky Mountain quaking aspen.[53] Or consider ancient individuals, such as the bristlecone pine (*Pinus aristata* and *P. longaeva*).[54] In 1964, Prometheus, one of the trees on the northeastern exposure of Wheeler Peak, was cut down and sectioned by scientists for the purpose of counting its annual rings. The tree had lived for more than 4,900 years, according to the National Park Service's Web site.[55] Today, the oldest living bristlecone is thought to be about 4,600 years old, according to the same Web site, and lives in the White Mountains of California—may the Park Service save it from scientists! Other Methuselahs are found among Patagonian cypresses (*Fitzroya cupressoides*; 4,000 years) and the giant sequoia or Sierra redwoods (*Sequoiadendron giganteum*, also *Sequoia sempervirans)* of the western slopes of the Sierra Nevada Range (approaching 4,000 years).

In invertebrates, negligible senescence and indefinite life span seems to be allied with a potential for regeneration and a habit of vegetative reproduction (Placozoa, Porifera, Cnidaria, Platyhelminthes, Entoprocta, Ectoprocta, Annelida, and Urochordata, our sister chordate phylum[56]). A few vertebrates, especially females, also fail to grow old or die "naturally." Rather than settling down to a definitive adult size they grow continuously. Typically, indefinite growth is found in cold, deepwater scorpaenid rockfish, cod, iteroparous salmonids, halibut, and possibly perch and arctic char.[57] A lake sturgeon, weighing 215 pounds and measuring 81 inches in length, caught in Ontario in 1953, was estimated to be 152 years old on the basis of annual rings in its pectoral fin, although some gerontologists remain skeptical.[58] Negligible senescence and indefinite life span are also reported for some amphibians (for example, female *Xenopus*), box turtles (123 years),[59] painted turtles and Blanding's turtles,[60] the famously long-lived Marion's tortoise (*Testudo gigantea* or *Geochelone gigantean*) from the Seychelles (who died accidentally at something between 180 and 200 years of age in 1918 on Mauritius after 152 years in captivity), and the Galapagos tortoise (*Geochelone nigra abingdoni* also known as Lonesome George),[61] some snakes (for example, the water snake *Natrix natrix helvetica*),[62] pelagic birds,[63] the common tern and budgerigars.[64]

Finally, estimates of age in bowhead whales based on aspartic acid racemization of the eye lens range up to 211 years ± 35 years (standard error). A correlation between extreme size and extreme longevity are not unexpected. "The energetic and skeletal requirements of attaining great body mass demand an extended life-span simply because growing large takes time."[65] And the extreme body mass of bowheads would certainly be expected to place them at the high end of animal longevity. What is more, body mass is frequently inversely proportional to temperature, and these whales, living in the high-latitude polar seas should be long-living if only because of ambient temperature. But what is surprising is the absence of senescent changes in bowhead corpses, for example, the absence of cataracts in the lenses of aged adults and prolonged fecundity in most of the oldest females and males.

Many explanations based on adaptive advantage in the course of evolution are offered for the phenomenon of negligible senescence and indefinite life span of these plants and animals. In vertebrates, especially in mammals, life-history variables have several parallels in allometry (the correlation of growth of a part to the growth of a whole organism): longevity is directly proportional to body mass (especially brain mass in the case of human beings), birth mass, length of gestation, and age at sexual maturity while inversely proportional to number of offspring born at the same time, duration of lactation, and growth rate.[66] On the other hand, some long-lived plants profit from isolation and lack of competition, but they are also adapted to life in inhospitable terrains. And, clearly, resilience aids longevity: "the greater the number of defects the organism can accumulate while remaining alive, the greater its life span will be."[67]

Of course, like everything else in the real world, organisms will suffer many lesions—"the thousand cuts" that chip away at life[68]—until systems cease to function. And death by traumatic injury, especially for animals, can always come out of the blue. For example, the German bombs that destroyed parts of Edinburgh in World War II killed anemones that had lived there for eighty years without showing any evidence of aging.[69] Thus, the soma, as such, is not necessarily doomed, especially if it remains capable of growth and is able to avoid accidents and hazards.

LIFECYCLES, MORTALITY, AND IMMORTALITY

Contrary to Weismannian expectations, germ cells in both multicellular and unicellular organisms die. What is more, the somata of some multicellular organisms exhibit negligible senescence and indefinite life spans. "It is unlikely [therefore] . . . that understanding the immortality of the germ line will tell us much about the mortality of the soma" or vice versa.[70]

On the other hand, to whatever degree mortality, negligible senescence, and indefinite life span evolve in both unicellular and multicellular parts of the lifecycle, both have something to tell us. Neither part is immune and neither possesses a monopoly on mortality or negligible senescence and indefinite life span, but both play specific roles in the lifecycle: the germ line prepares cells for entering the somatic line and the somatic line prepares cells for entering the germ line. The male germ line may imprint genes with silencing instructions that facilitate contact between the blastocyst's outer layer of cells (trophectoderm) and maternal tissue, while the egg, or oocyte, to be precise, can reprogram the genome (the species' census of genes) and make the full spectrum of developmental events available to the cells of the developing multicellular organisms.

The oocyte's ability to reprogram a genome in preparation for development extends to foreign nuclei artificially introduced into eggs. This is the lesson of cloning. Indeed, at a time when most biologists were caught sleeping, Gina Kolata, the science writer, had the perspicacity to appreciate the centrality of the egg's talent for reprogramming and the potential, hence, of cloning: "[If] scientists . . . could learn how the egg reprograms a cell's DNA, bringing it back to its primordial state, they might someday be able to force a cell to reprogram its *own* DNA and then differentiate into any sort of cell that the scientists want. That, of course is the most futuristic scenario of all, . . . but it shows what might someday be possible. That process of learning to reprogram a cell's DNA would have to begin, however, with cloning."[71]

The remarkable thing about reprogramming is that it allows cells produced by cell division to do more than their parent cells were capable of doing. In the lexicon of modern biology, such cells are said to be "pluripotent." When a nucleus acquires pluripotency, for example, as a consequence of nuclear transplantation to an oocyte, the nucleus acquires the ability to direct differentiation into a variety of—if not all—cell types present in the body. Indeed, it is because nuclear transfer to oocytes would seem to allow us to make pluripotent cells that the process holds therapeutic promise.

But pluripotent cells are also produced in the body of the embryo and fetus. When transferred and raised in tissue culture, these pluripotent cells are called *embryonic stem (ES)* and *germ stem (GS) cells*, while the "genuine parts" in the adult organism are thought to give rise to the *precursor* or *transit amplifying cells* that proliferate and produce the cells that differentiate into tissues at every stage of the lifecycle. Organisms, such as hydra and planaria, able to support massive amounts of regeneration, have stem cells "akin to germ cells in that they maintain the continuity of a line of descent from generation to generation,"[72] albeit remaining somatic in type.

Thus, as life cycles, lifecycles move from germ cells capable of reprogramming to stem cells capable of proliferating and initiating the programs of precursor and transit cells, interacting with other tissue and with the external environment, and differentiating in tissues and organs. Mortality represents an end of these processes; negligible senescence and indefinite life span represent their continuation.

The Germ Line in Mammals Generally and
Homo sapiens Specifically

Typically, in the healthy mammalian soma, large-scale cellular death is accompanied by large-scale cellular proliferation, especially among transit amplifying cells in bone marrow, epidermis, and intestinal lining epithelium (see chapter 5). Likewise, the male germ line in the testis is maintained through high rates of proliferation in transit spermatogonial cells, while in the female germ line of mammals, until recently, cell division in oogonial transit cells was thought to have been completed during fetal development.[73]

Female and male germ cells are hardly comparable except that their nuclei undergo meiosis during a so-called -cyte stage, "oocytes" in the female and "spermatocytes" in the male. In female mammals, oocytes first appear in the fetus and are by far the most conspicuous germ cells, if not the only representatives of the germ line present at birth, while in the male, spermatocytes do not appear until puberty, and some stem and precursor spermatogonia go on to become spermatocytes throughout the adult lifetime.[74] Germ cells are continuously lost, however, in the female typically via atresia or the failure of follicles to perforate and release mature ova, and in the male by massive waves of cell death and by ejaculation. But whereas germ-cell loss in female mammals remains largely or completely uncompensated, in the male, lost germ cells are replaced by division throughout the postpubescent years.

The female germ-cell population in human beings reaches its peak population of about 6,000,000 at week twenty of gestation and shrinks drastically in waves of cell death to something between 2,000,000 and 800,000 at birth, and from there to 40,000 or less at puberty. In an average woman, only 400 or so oocytes mature completely and are ovulated during a lifetime (that is, about 40 years at 10 oocytes per year, assuming time off during pregnancies and lactation). Only mature or ripe eggs are capable of productive fertilization[75] and supporting complete development.

The population of oocytes continues to plummet at each menstrual cycle, leaving only 27,000 oocytes at age 37. Thereafter, oocytes are lost at an even faster rate. At 50 ± 7 to 8 years, the ovary is bereft of all but 1,000 oocytes, if that.

In female rodents, germ-cell wastage reduces the oocyte population from 6.4×10^4 oocytes at day 17.5 of pregnancy, to 1.9×10^4 oocytes shortly after birth. But the rodent ovary manages to regenerate some of the lost oocytes.[76]

The cause of all this death in female germ cells remains uncertain, but hormones are the chief suspects.[77] Alternatively, the death of individual oocytes may be "self-inflicted" and an adaptation for solving the problem of accumulated mutations. In this event, the chief suspect switches to mitochondria, the famous mighty mites of cellular metabolism, or energy factories, but

	fetus	newborn	puberty	adult	senescent
female					
number of precursors	many	maybe some	maybe some	maybe some	0
number of oocytes	6,000,000	1,00,000–2,000,000	40,000	~27,000	1000 to 0
male					
number of precursors	some dormant	some dormant	many	many	many
number of spermatazoa	0	0	very many	very many	many but not as many as before and not as healthy

FIGURE 3.6. Numbers of precursors, oocytes, and spermatazoa in *Homo sapiens.*

also the source of a death-inducing message triggering programmed cell death. Indeed, the death rate of oocytes is retarded when fresh mitochondria from follicle cells are microinjected into oocytes.[78]

In contrast, the production of spermatozoa continues from puberty throughout the lifetime, although, in many mammals, it is interrupted seasonally. Cell renewal through division keeps up the pace of spermiogenesis or spermatozoan differentiation, and the male *Homo sapiens* may remain fertile into his dotage, although the spermatozoa may deteriorate in quality, not "swimming" as well or quite as competently to fertilize eggs and, more importantly, failing to support healthy development.[79] Sperm counts in ejaculates may also decline modestly with age. Germ-cell wastage is, however, even more rampant than in the female. Differentiating spermatozoa are phagocytized by sustentacular cells and testicular macrophages in seminiferous tubules, and spermatozoa that survive the damage inflicted by sheering force during ejaculation die in droves in the female genital track or elsewhere, as the case may be.

IN SUM

The cyclic and linear arrow models of life are starkly contrasting: while lifecycles offer no obstacles to the evolution of death, the linear or Weismann/Haeckel model provides an immortal germ and a mortal soma but no room for death's evolution. At the crux of the issue is whether the lifecycle itself is a fundamental part of life, subject to evolutionary pressures. If lifecycles evolve, then the rate at which living things, including somata, move through their life-cycles should be up for grabs, so to speak, and neither mortality nor immortality should be ultimately constrained or confined to one or another part of the lifecycle. The death of germ cells and the extension of lifetimes through negligible senescence would seem compatible with the possibility of lifecycle evolution and even the possibility of relative immortality.

The sexual lifecycle of large animals, such as us, transforms the parts of virtual offspring and parents to haploid and diploid parts, the latter containing conventional stages marking life's passage—from fertilized egg, preembryo or blastocyst, embryo, fetus, neonate, juvenile, adolescent, adult—with meiotic reduction representing a transition on the way to eggs and spermatozoa. Some eukaryotic lifecycles are short, but all would seem plastic and susceptible to evolutionary change.

Eukaryotic animals as such do not pass through an entire lifecycle. Complete passage depends on moving from a multicellular, diploid part with a variety of stages to a single-celled, haploid part comprised of differentiated but nondividing germ cells. What is more, the lifecycle can be cut short in any part, phase, period, or stage by death. On the other hand, theoretically, nothing would seem to prevent the large, multicellular form from reinforcing its own longevity by providing itself with the resources of prolonged life, and, indeed, organisms with indefinite life spans and negligible senescence seem to have done just that.

In contrast, the linear Weismann/Haeckel arrow requires every mortal multicellular organism's lifetime to begin with an ancestral, single cell—derived from the immortal germ line—to develop into a soma through the truncated and progressive recapitulation of an evolutionary history. Weismann/Haeckel broke up lifecycles and confounded their theoretical wave function at each generation. They recognized the difference between the body or soma and sex cells or germ, but mortality—with aging, senescence, and dying—was attributed exclusively to the soma. The adult body or soma of multicellular organisms is said to act "merely as a host for the perpetuation of immortal germinal material,"[80] evolving into large and complex multicellular organisms containing differentiated cells of numerous types, tissues, organs, and organ systems. The germ, on the other hand, is sequestered in its pristine, ancient, and uncontaminated, unicellular condition. Theoretically, the division of labor between a nourishing soma and a reproductive germ determines their different directions of evolution and development. Natural selection doomed, the soma to mortality as a consequence of differentiation, while, in the absence of this differentiation, the unicellular germ remained immortal.

The difference between the cyclic and arrow versions of life centers on the contrast between immortality and mortality. From Weismann/Haeckel's point of view, the soma cannot be immortal, while from the lifecycle point of view, nothing, other than compromises with other adaptations and sheer bad luck, would seem to prevent the evolution of death and the soma surviving indefinitely.

Chapter 4

Keeping Life Afloat

[T]he determination of degrees of longevity and of the fact of death itself, is inherent in the innate, hereditarily determined biological constitution of the individual and the species.

—Raymond Pearl, *The Biology of Death*

"What about death?" she said. . . .
"This is one of the basic questions of our time," he said. "If we knew how to make a good job of death, it wouldn't be so frightful, would it? The famous prizefighter Joe Louis has been quoted as saying that everybody wants to go to heaven but nobody wants to die. I've used that in many of my sermons."

—Don DeLillo, *Americana*

The great irony exposed in chapter 3 is that death contributes to life principally by providing the sink for life's waste—corpses. Chapter 4 examines some of death's other effects on life, effects that turn out to be salubrious rather than corrosive as might have been expected.

The chapter begins by answering the question, why is life so profligate? The analysis moves from virtual life, far from equilibrium thermodynamics, to down-to-earth life in the gambling halls of probability. The examination then moves to life's mechanistic side, suggesting how gaps created by death in the fabric of life are filled with fresh material provided by life, how life is constantly buoyed up and paid dividends on its surplus. Chapter 4, thus, confronts the concrete issue of death's payoff for life, the advantages death offers for life and how evolution may trump those advantages.

WHY IS LIFE SO PROFLIGATE?

Please consider the relatively simple task of copying a picture, for example, a photograph of a family member. I choose this task because copying suggests reproduction, but I have more general phenomena in mind: replacing, replenishing, restoring, and reproducing life's parts as well as living things. Despite having access to the very best equipment and competent personnel, the copy lacks something. It may be good enough, but, without fudging, it is never quite as sharp and vivid as the original. What is more, were one to make copies of copies in succession, the image would deteriorate further. The subject might even become unrecognizable. Actually, this result is predicted by information theory's version of the second law of thermodynamics: some information is lost whenever information is replicated.

Now, compare copying a picture to the crucial task of replicating DNA in a cell. Information theory predicts that every time the cell divides it loses information, and, multiplied by billions of cells dividing, the loss of information to the organism could become catastrophic. In fact, life invests vast amounts of energy and resources in proofreading and editing, thereby preventing this loss of information, but, when all else fails, death steps in and removes flawed cells. Call it cellular senescence or tumor suppression, but the fact is that cellular death is the best anticancer agent working for us! Cellular death plays the role of a failsafe mechanism, and its intervention probably prevents the development of many tumors before they spread their destructive blight in the organism.

But cellular death plays a variety of additional roles in life. For example, life-support systems of many differentiated cells, such as blood cells and stomach and intestinal epithelia, are so utterly compromised that the cells' life is unsupportable. And death plays a role, if not the apotheosis of differentiation for other cells. For example, keratinized keratinocytes of the epidermis create a barrier to osmosis, and hence dehydration, in the very act of dying. Thus, death not only steps in and removes the effete cell without placing an additional burden on living cells, but utilizes the dead cell in an otherwise unobtainable physiological role.

Probably the most dramatic appearance of death in a salubrious role occurs during metamorphic transitions between stages of a lifecycle. The metamorphosis of a tadpole to a frog or an insect imago to a pupa involves the death of massive amounts of larval tissue virtually simultaneously with replacement by adult tissue.[1] But death also plays a morphogenic, if less dramatic, role during the sculpturing of vertebrate bodies, for example, freeing the limb from the body wall and removing webbing between digits.[2]

Death may come to the rescue of life in several ways, but why does life cycle in the first place? Why doesn't it just sit still? The answer is well known by biologists: living things are *dissipative structures* that draw upon and dissipate external energy as they develop, maintain, and replace their ordered and organizing parts.

One is aware of some of this exchange when one eats and exercises. Dieting is, of course, a favorite pastime, but starvation and malnutrition are also known to shape the organism. Severe disease, such as scurvy, may even result from a failure to maintain a healthy diet. Weight-bearing tension and pressure also alter the dynamics of bone deposition as well as muscle tonus and strength.

Thus the regulation of exchange may be undertaken through conscious effort (for example, exercise and diet intended to maintain a healthy cardiovascular system). But most daily turnover occurs in places and ways unknown to us. In fact, "we replace many cells every 6 months and most of our body every seven years, so that we truly become what we eat many times during our lifetime."[3] In other words, we keep ourselves alive by cycling materials through us—in and out of ourselves. This funneling of energy and material means, in thermodynamic terms, that we are open systems, even if we tend to think of ourselves as self-contained entities and even if we forget that death is part of this recycling. Indeed, death is the heart of recycling!

LIFE IN OPEN SYSTEMS

Life cycles through open systems. In fact, living things constantly draw energy and materials from outside their confines and utilize these resources in the creation and re-creation of their complex, coordinated, and highly differentiated structure. Life is *so* open that one must include nothing less than the solar system before one encompasses all the sources of energy and material utilized by living things on Earth! The sun's radiant energy furnishes most of the free energy utilized on Earth, which is to say, powers the photosynthesis that powers most of life as we know it. But "most" is not all, and some widespread, if offbeat, living things depend on non-solar energy sources—sulfur-eating bacteria, methane-eating archaea—whose effects trickle up through the Earth's overall economy.

In addition, the atmosphere stores radiant energy and maintains the ambient temperature where physiological reactions take place. The Earth's molten core also provides internally generated energy affecting the mantle and manifest in plate tectonics, devastating volcanism, and earthquakes, and leading to mass extinctions and allopatric speciation. Thus life on Earth would not be what it is without solar energy, the atmosphere, the Earth's molten core, and on and on.[4]

Life would not be as we know it on Earth without the moon in its tidal orbit stabilizing the Earth's axial tilt, moderating climate and seasons, imparting tides, weakening wind, and lengthening or shortening days.[5] What is more, there would probably be no life on Earth whatsoever were Jupiter not sweeping hordes of asteroids from the inner solar system and playing "the role of peacekeeper, deflecting most of these dangerous leftovers from the solar system's creation into its far reaches, where they present little danger" to us.[6] Indeed, the last major impact from a large asteroid some sixty-five million years ago contributed to one of Earth's mass extinctions and probably set the stage for the emergence of mammals as the major form of large animal on the contemporary Earth. Finally, or initially, as the case may be, one would have to go back to the solar system's beginning more than four and a half billion years ago in order to include the kind of events that made Earth a planet capable of supporting life. Then, one might have an idea of life's openness and the scope of the systems through which life cycles.

How then does death enter the calculus of open systems feeding lifecycles? Most conspicuously! In terms of trophic relationships, especially predator/prey relationships, death plays the role of making prey available as a source of material and energy. Were it not for death, everything we eat would be a potential parasite! Beyond relationships with other living things, death is the final phase in the transition of material and energy passing into living things. Death occurs whenever one living system is drained into another. This passage is not metempsychosis or transcendence. This passage is quintessential movement through life's open systems.

Life Takes Place Far from Equilibrium

Biologists also know that life, or dissipative systems generally, do not cycle close to thermodynamic equilibrium. Unlike nonliving things residing at or near thermodynamic equilibrium, living structures cycle in a domain *far from equilibrium* where they have unique properties, especially the spontaneous origin or creation of apparent patterns.

A formal (and respectably reductionist) conception of far-from-equilibrium thermodynamics was first devised by the Belgian theoretical chemist Ilya Prigogine (1917–2003) while advancing research by the Norwegian-American chemical physicist Lars Onsager (1903–1976) on irreversible process. According to Prigogine, living things (but not only living things) provide a "striking example of the fundamental new properties that matter acquires . . . [far from equilibrium and] can be 'perceived' by the system, creating the possibility of pattern selection"[7] and, hence, natural selection.

As life emerges far from thermodynamic equilibrium, symmetry is broken, dimensions are created de novo, and patterns appear. Initially, oscillations become frequencies and length arises by splitting (bifurcation). Farther away from equilibrium, simple periodic behavior becomes complex aperiodic behavior; both stochastic and deterministic events appear in the same system; stable and unstable behaviors unfold simultaneously; necessity and chance meet; and living structures ultimately materialize as a "result of self-organization . . . [even if] we must admit that we remain far from any quantitative theory."[8]

Finally, far from equilibrium, control (constraints and regulation) and change (evolution and development) come from things as themselves: Mathematical rules determine biological shapes and processes; stable patterns control their own survival; a change in any part is likely to produce a reaction in many other dependent parts of the whole; the product of one iteration feeds the next iteration; origins occur slowly, while diversification emerges rapidly; creativity peaks and diverse forms fall to extinction.

This is the zone of *self-organization.* "Where large numbers of individuals act simultaneously, a system can suddenly break out of an amorphous state and begin to exhibit order and pattern. . . . In place of explicitly coding for a pattern by means of a blueprint or recipe, self-organized pattern-formation relies on positive feedback, negative feedback, and a dynamic system involving large numbers of actions and interactions."[9] But, at the same time that physical life emerges far from equilibrium, *virtual life* is taking place at the edge of *chaos.*[10] "Chaos seems to enable areas of greater order to be fenced off as life inside the general thermodynamic trend to greater disorder."[11] Indeed, virtual life is the chaos detected amidst the apparent order resulting from repetition and the dampening effect of negative feedback. Chaos is the nonperiodicity and nonlinearity—repetition without a constant interval and repetition with a difference—which is to say, life's unpredictability.[12]

Ironically, the problem of ascribing chaos to life is, in part, due to the word "chaos," which implies disorder rather than a type of order not covered by the usual definitions of stability, stasis, periodicity, and return. When the mathematician James Yorke coined the term,[13] "chaos" was intended to be "mysterious and mischievous,"[14] which does not help one understand it. In order to avoid confusion, one must dissociate chaos from randomness, especially regarding the lowest free energy state of equilibrium.

Freed of randomness, life and chaos become partners in Stuart Kauffman's concept of emerging properties and his science of complexity,[15] and in Humberto Maturana's and Francisco Varela's notions of self-organization or autopoiesis—the self-producing organization allowing an organism to pull itself "up by its own bootstraps and . . . [rescue itself] from its environment through its own dynamics."[16] Thus chaos is at the center of oscillatory functions, such as the periodic release of insulin and other hormones, and at the crux of "dynamical disease"—disease resulting from the failure of normal periodicities—such as myocardial fibrillation. And, through chaotic dynamics, life leaps into the novel configurations appearing irregularly in evolution.

Dying—when living things become corpses—traverses the distance between far-from-equilibrium and near-equilibrium thermodynamics. But corpses and fossils are "hard copy" reminders that "chaos—deterministic and patterned—pulls the data into visible shapes."[17] In the world of chaos far from equilibrium, one does not put one's finger on living things. Rather, virtual life proceeds in living things and virtual death makes corpses from living things. This is where the game of life is played, and life, like chaos, "is really unpredictable!"[18]

Life's Strange Attractors: Life among the Fractals

With the change in perspective brought about by moving from near equilibrium to far from equilibrium, living things also move away from irreversible, thermodynamic processes toward reversible chaotic processes. This is the domain of *strange attractors,*[19] "islands of stability,"[20] where the "transition to chaos is complete,"[21] where life acquires all its many forms as it becomes one with the virtual.

Unlike most objects in the nonliving, thermodynamic universe,[22] strange attractors exhibit nonlinear dynamics, various instabilities, exquisite sensitivity to initial conditions, and conspicuous fluctuations. In other words, strange attractors give the appearance of being alive! Moreover, strange attractors contain substructures that are also strange attractors, which is to say, strange attractors are "fractal": composites of self-similar entities scaled and repeated in space and time. Indeed, a fractal structure seems to be a very ancient feature of living things[23] still epitomized by the modular structure of organs within organ systems, of tissues within organs, of cells within tissues, and even DNA within cells.

Theoretically, living things may remain near-periodic residents in the phase space of strange attractors or escape into the terrain of irresolvability. They may then enter the phase space of another strange attractor and acquire a new near-periodic function, and hence evolve, or they may never resolve into another strange attractor and vanish into virtual extinction.

With their death or extinction, living things fall to the state of nonliving objects and energy ever-approaching thermodynamic equilibrium, namely to the state of corpses, fossils, decay, and dust. But, from the perspective of virtual life, of chaos and strange attractors, virtual death is not a fall so much as falling, not an effect but something effecting. From this perspective, death confines lifecycles to the phase space of strange attractors and absorbs life's endless repetition within fractals' endless expansion.

GAMBLING ON LIFE: DEATH AGAINST THE ODDS

Moving down from the Promethean hills far from equilibrium, death enters the more familiar, if no less abstract, valley of probability and life's gambling casinos, where death may be the end point in a game of chance. Many of us will imagine that one has some control over life and death, and indeed one has. To a degree we control what we eat and how often, when and how long we sleep, what we do and how hard we work, how vigorously we exercise, how often we have sex, and so on. In academia, it's still "publish or perish" and elsewhere it's "sink or swim," but from the point of view of statisticians (which is to say, gamblers), life is more nearly a matter of luck. "*Random events . . . events that are unpredictable except in terms of probability*"[24] hold life in the balance. For gamblers, death is losing a game of chance.[25]

Chance can intrude at any point in a lifecycle merely by the luck of the draw or the fall of the die (pun intended). Death can result from accidents of birth—the genes we inherit—and accidents along the highway of life—literally! What is more, the benefits acquired or the damage inflicted in an early stage of life (e.g., an athlete's heart with an enormous stroke volume or the infectious disease we had as a child) may kick in to kill us at a later stage.[26]

A gambler may believe that good or bad luck can lead to fortune or misfortune, in contrast to the statistician who relies on the odds, but both gambler and statistician approach the playing table of life with the same combination of stochastic pragmatism and stoical acceptance. When one enters a gaming casino, one assumes the games are honestly based on randomness, while the odds or probabilities of winning any game are set by the house and always slanted in its favor.[27] In the long run, therefore, the casino will make money, but one can still prevail if one takes one's winnings when one has them. Life would seem much the same.

WHAT ARE THE ODDS OF LIFE AND HOW ARE THEY SET?

The probability of winning at roulette or at any honest gambling table is calculable by the simple devise of converting frequencies to probabilities. The probability of winning at life is more difficult to calculate, since frequencies of successful living are obtained only after the fact and would seem to change with the times, place, history, background, gender, and so on.

Indeed, the odds of one's living longer have improved between 1900 and 1999, and individuals born in the twenty-first century are likely to live even longer than individuals born in the twentieth century.[28] The preponderance of gerontologists seem to be of the opinion that this change in odds is due to improvements in the environment: in public health, sanitation and hygiene, the quality of air, water and food, malaria and yellow fever control, smallpox and polio eradication, medicine generally and especially the control of infectious disease, wealth, occupational and domestic safety, reduced hazards in industrial and agricultural work, education, and modifications in behavior (such as the increased acceptance of condoms and the decline in smoking[29]). But the odds on living are not constant or homogeneous for people ostensibly sharing the same environment, as they might be were the odds set merely by the environment.

Data on age at death show that females survive longer than males at every age. For example, data from Singapore (chosen for completeness over a ten-year period and without other criteria for selection from the World Health Organization mortality database[30]) show that far more women (F) live beyond the age of seventy-five years than men (M). Indeed, the number of women living beyond the age of sixty-five years is about the same as the number of men surviving to sixty-five years. As a matter of fact, the inequality between men and women runs to far earlier ages. On average, for every year throughout the decade, roughly 10 men die before the age of fifty-five for every 6 women.

Even among infants, where phenomenal progress has been made at reducing mortality, males suffer a disadvantage compared to females (in each year's data, the bar for M is higher than the bar for F). As the decade of the 1990s proceeded and the odds on infants surviving the first year of life generally improved, male infants still tended to die in substantially higher numbers than female infants.

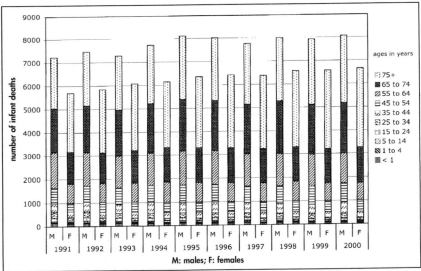

FIGURE 4.1. Total deaths in Singapore, 1991–2000. World Health Organization Mortality Database, 2004.

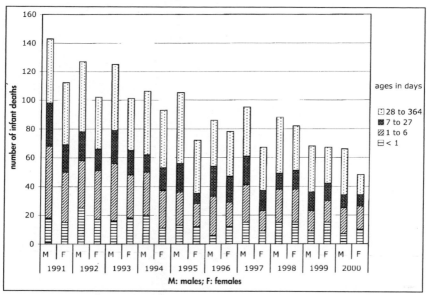

FIGURE 4.2. Infant mortality rates in Singapore, 1991–2000. World Health Organization Mortality Database, 2004.

What these data say is that the odds are stacked in favor of females of the species. When it comes to gender, life is not a crapshoot or a poker game where everything is contingent. The genders are not equal!

IS LIFE A FAIR GAME?

Ignoring gender differences, if life were a fair game one would expect the chance of death to be the same or virtually constant for individuals at every age following birth. Sorry! For most of our post-adolescent lifetime, the odds are increasingly stacked against us, and the likelihood of reaching succeeding birthdays falls from year to year until old age.

Formally, life expectancy (sometimes called "life-after x," where x is attained years) is the number of years an average person can expect to live having once reached a particular age. In practice, life expectancy is a computed, hypothetical value for a cohort (people born the same year) calculated from trends in mortality over several years. In figure 4.3, life expectancies are plotted for two cohorts of all races and both genders born in the United States: a cohort born in 1901 (filled squares) and a cohort born in 1999 (open diamonds). The figure shows, for example, that people born in 1901 and reaching their early twenties could expect to live another forty-some years and die

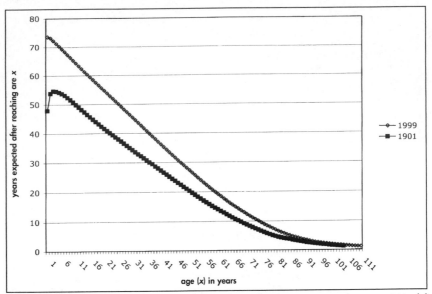

FIGURE 4.3. Life expectancy as a function of age in attained years. World Health Organization Mortality Lifetables USA 1901–1999.

in their early sixties, while people born in 1999 and reaching their early twenties can expect to live another fifty-five years and die in their late seventies.

The initial portion of the curve for the 1901 cohort (the neonatal period) indicates that odds for living actually increase after birth! The large difference in the early portion of the curves would seem to reflect improvements in standards of living between 1901 and 1999, especially in the quality of medical care and nutrition available to average citizens. Infectious diseases would have killed more individuals early in life in the 1901 cohort than in the 1999 cohort, and the general health, especially of pregnant women, would have played a large role in determining the initial shape of the curves as well. Interestingly, a recently reported sharp rise in longevity among the elderly in the late twentieth century may reflect a delayed response to improvements in the treatment of infectious disease in early life.[31]

A period of high life expectancy corresponding to the juvenile stage is followed by a change in slope, and the life expectancy curves begin their characteristic, exponential decline. The curves' downward trend reflects increasing, age-specific mortality. In fact, death or mortality rates double every seven to eight years over most of a lifetime. This trend was discovered by the English actuary Benjamin Gompertz (1825–1865) and was later refined by William Makeham (1812–1884). It is known as the Gompertzian (or Gompertzian-Makeham) exponential increase. As one lives through the greater portion of a lifetime, one's reasonable expectation of living longer is rapidly whittled down.[32] Gavrilova and Gavrilov argue that this "age-dependent component of mortality is determined not by social conditions, but by significantly more stable biological characteristics of human populations."[33]

At the end of the Gompertzian increase, when death rates no longer increase, the number of years an average elderly person can expect to live settles down, but at very low life expectancy (flat portions of curves for both cohorts). Remarkably, both curves level off at about the same life expectancy where they become overlapping. In other words, at the end, the game of life is played on a level playing field, and the probability of living an additional year remains more or less unchanged.

LEVELING THE PLAYING FIELD

Empirical data on supercentenarians (people 110 years and older) support the contention of a "late-life mortality deceleration," "mortality leveling-off" at advanced ages, or the "late-life mortality plateau,"[34] to more or less even odds for living another year at extreme old age. Using his data set of 696 verified supercentenarians, the gerontologist Robert Young estimates that "the death rate for ages 110–113 would be between 50% and 55%,"[35] and the gerontologist L. Stephen Coles, using the historically validated Worldwide Supercentenarians' database estimated that the mortality rate for supercentenarians is ~1/2, that is, "a correct flip of a fair coin is what it takes to migrate to the next year or fall into the arms of the Grim Reaper."[36] Thus, about half the members of this exclusive group of supercentenarians will die in any year, but, barring statistical quirks, the odds of dying are relatively constant compared to the increasing odds of dying during earlier adult life.

Thus, two related conclusions are drawn from the overlapping portion of the life expectancy curves: (1) Everyone will eventually die under the sword of probability, and (2) human life, as such, has no biological limit. To date, only Jeanne Calment has beaten the odds to the age of 122+ years,[37] but whether anyone will live longer is only a question of anyone living that long in the first place. After reaching 122, the odds of reaching 123 are even.

These conclusions are not actually too surprising. In fact, the notion of a Gompertzian increase continuing into old age has little support in animal studies. James Carey concludes from his medfly studies that "slowing of mortality rates at advanced ages in all studies suggest . . . that it is not possible to specify a specific life-span limited to the medfly and, by implication, to that of any species."[38] And Gavrilov and Gavrilova declare that "the observed dependence of mortality on age does not support the hypothesis that there is a species-specific life span limit."[39] Of course, the likelihood of death is, no doubt, complicated by numerous mechanisms affecting mortality, including repair mechanisms at old age compensating for damage at young age, and selection at old age for individuals with lower death rates, but, in the absence of life span limits, life in old age is truly a toss-up game of chance.

But what does one make of the convergence of the two life expectancy curves in the figure if human longevity has no maximum and life span no limit? Clearly, similar dampening of death rates takes place at about the same age in the two cohorts. Indeed, Gavrilov and Gavrilova have derived a function they call the "compensation effect of mortality" that remains constant and toward which all contemporary populations gravitate. The compensation effect is "invariant relative to the living conditions and genetic characteristics of the populations under comparison. . . . For *Homo sapiens,* this quantity is 95±2 years."[40]

Life, thus, is not a fair game for most of a lifetime. The odds of surviving change throughout a lifetime, increasing briefly at first and then decreasing until they level off in old age. In the early years, life expectancy would seem very sensitive to local conditions, such as the quality of nutrition and the availability of medical treatment, while in the later years, life expectancy would seem virtually dependent on chance. Remarkably, the large range of life expectancies found during the period of Gompertzian increase would seem consistent with the unpredictable behavior of a chaotic population, since sensitivity to initial conditions is diagnostic of chaos. In contrast, the dampening of the curves toward an asymptote at extreme old age would seem consistent with stochastic behavior in a random population. Aging may anticipate the descent from far-from-equilibrium to near-equilibrium thermodynamics, unobtrusively converting life's unpredictability into probability.

IMPROVING PROFIT MARGINS

Gambling is not, of course, the only game of life in town. Business is just as sensitive to initial conditions and has many lifelike models to guide it. In fact, the body's requirement for cellular replacements is analogous to a retail business's requirement for merchandise. The organism is analogous to the business proper, the shop on Main Street or the showroom in the mall, while the organism's environment is analogous to the buying public or customers. (One might extend the notion of the shop to include upkeep, rent, personnel, and so on, and the environment to include location, the proximity of amenities, nearby bus stops, parking, and so on but those complexities are ignored in this basic model.)

A business may seem to exist for the purpose of meeting the demands of consumers, but in a free-market economy the business is devoted to meeting the demands of its owner(s) for profits. The success of a business depends on its operating at a profit. Likewise, in community ecology (which shares the same Greek root, *oik-* "house" or "habitation" with "economy"), the evolutionary success of an organism—better known as its fitness—depends on the organism reproducing or promoting the survival and reproduction of its descendents or those of its close relatives.

According to notions of supply-sided economics, the marketplace functions because supply creates demand, and the profits of business have a way of trickling down to consumer and community in general. Likewise, healthy organisms develop and persist, because what is good for the organism has a way of trickling down. In the dynamics of life, many organisms both develop and maintain themselves by renewing their cells against an endless tide of cell loss to the environment. While shedding differentiated and frequently dead cells, organisms resupply their "shelves" (that is, tissues) with cells freshly produced by cell division. Of course, business depends on merchandise arriving from external (and increasingly remote) sources, and organisms depend on raw materials arriving from outside sources, but the goods resupplying the organism's tissues, namely cells, are normally produced internally through cell division.[41] This process of normal cellular turnover is the business as usual of cellular retailing.

A SIMPLE BUSINESS MODEL

Life is profitable when, at the end of the day, the organism has a surfeit of cells, and life is bankrupt when it has run out of cells. In the embryo, cellular additions play the primary role in growth and development; in the adult, cellular substitution keeps the body in business; and during aging, cellular replacement gradually fails. But evolution has equipped us with a plethora of replacement cells with which to face the vicissitudes of life, and human beings, like most other animals, only die when effete cells are no longer replaced with normal, functioning cells.

Cellular replacement is not always a matter of replacement in kind, however. Following trauma, replacement may represent the difference between life and death. Cellular substitution may take the form of scar formation, in which dead cells in damaged areas are replaced by scar-forming cells that keep physiological losses in check but do not return the tissue to normal structure and function. Alternatively, scar formation may be prevented when regeneration takes over and dead or missing cells are replaced with new cells in a normally functioning structure.

But, in general, in the well-functioning adult, recruiting and mobilizing cells identical to or close relatives of the original cells is tantamount to self-maintenance and good health. Thus, in the adult, cellular replacement in kind is the rule. Indeed, the normal body renews billions of cells daily—conspicuously, blood, outer skin, and gut-lining cells—and the failure to replace these cells is the cause of bodily breakdown. Losing the capacity to replace the body's cells brings on disability, morbidity, and mortality. We suffer from defects and injuries; we get sick and become frail; and we weaken and die from many causes, but *the root cause is not having enough of the right kind of cells in the right place at the right time.*

Inevitably, many cells in the body may be expected to "go bad" quite normally. Indeed, cell death is built into many of the most important mechanisms of cell differentiation, for example, the keratinization of the epidermis or outer skin. In addition, many blood cells and tissue macrophages normally die in the course of fighting infections and coping with trauma. But cells also get wasted by the environment: allegedly by mercury and phthalate esters in food, dioxins, lead and asbestos in the surroundings, smoke in the atmosphere, radiation from natural sources, and free radicals from normal metabolism. Cell toxins may damage cells or accumulate and interfere with normal cellular activities, or they may thwart reliable cell division and lead to the synthesis of faulty proteins with impaired function. But, for the most part, the organism compensates by replacing spent, worn-out, and damaged cells with healthy new ones.

Irreparable damage is only done when replacement cells are no longer able to maintain normal cellular activities or restore impaired functions. Many examples are readily cited. Acute inflammation may kill by permitting or inducing cancerous change in replacement cells.[42] Cancers kill because they invade tissues and destroy normal cells while preventing replacement. Infections kill because they destroy cells or impair their ability to replace damaged cells quickly enough to maintain bodily functions, for example, retaining bodily fluid in the case of cholera or radiation poisoning. And coronary disease kills because ischemia or necrotic tissue interrupts the heart's normal, rhythmic contractile impulse or because scar tissue produced by fibroblasts has replaced heart muscle cells (cardiac myocytes) and impaired contraction. Finally, if we outlive all the hazards of "natural life," *we suffer from aging and "natural death" because we have run out of the normal replacement cells that previously maintained our body's working tissues.*

OTHER BUSINESS MODELS

Management must keep track of inventory and minimize warehousing, while replacing sold merchandise quickly and efficiently. Likewise, evolutionary history prepares a species to pay the premium for warehousing in the event that resources are undependable, and evolution strips organisms of backup in the event that high-quality resources are in ready supply. Thus, organisms may keep reserve cells in some tissues where they may be required to meet the demands of trauma and disease on short notice, while self-renewing stem cells may keep other tissues supplied with a constant stream of amplifying transit cells.

Accidents, of course, will always happen, and contingency plans for dealing with them must also be part of a successful business model. Disasters, such as a fire on the premises, can be met by a fire sale intended to get rid of damaged goods and make way for fresh merchandise. Likewise, a body damaged by wounds or disease may divert resources from one function to keep the organisms in business.

In practice, various strategies for profitability are employed in business and life. The shop that is constantly holding a going-out-of-business sale is making a profit by selling off consignment merchandise without regard to continuity. Some organisms follow this model, swarming when resources are available and virtually disappearing, or wintering over, when resources are scarce. This strategy would seem to have been adopted by rotifers and roundworms, for example, which have very limited numbers of cells but a system of parthenogenesis or hermaphroditic/male reproduction that takes advantage of opportunities for reproduction as quickly as they arise. Beginning with a parthenogenetic or self-fertilized egg, rapid cell division produces determined cells capable of creating a larva and reproducing a new generation in virtually no time.

Another business model is employed by large, established businesses, and by expanding businesses: constant resupply or additions of new merchandise to shelves or pallets, as the case may be, as holes are left by customers consuming previously available merchandise. Organisms with negligible senescence and indefinite life spans would seem to match cell loss with cell replacement, while organisms with indefinite growth would seem to add some cells, while, at the same time, replacing others.

In familiar organisms, including us, cellular supplies or inventories of cyclic goods are not maintained indefinitely. Life would seem to be held in a more delicate balance and built in obsolescence would seem to operate, ultimately undoing the organism. In organisms that age and die, cells have only so much shelf-life. After a while, cells cease dividing and renewing. A cell line that has thus lost the potential for resupply, peters out as its "shelves" (tissues) are emptied and laid bare. Like a business without stock and a capacity for restoring its inventory, an organism bereft of dividing cells forfeits its profits and fails—dies.

According to this cellular theory of aging, death results when the supply side of cellular dynamics fails and the ability of cells to divide and supply new cells is exhausted. Death may be precipitated by a host of circumstances that place extraordinary demands on the body for cellular resupply, but the death of the organism in old age is the consequence of body cells' failure to renew themselves.

On the other hand, life would seem viable as long as stem-cell supplies (discussed at length in chapter 5) remain in concert, supplying tissues with requisite cells as a whole. In the evolution of death, thus, greater coordination among stem-cell supplies would seem to prolong healthy longevity. *Death may evolve, therefore, through the accumulation and more efficient use of stem cells.*

IN SUM

Death supports life in a variety of ways, from fanning the flames of far-from-equilibrium thermodynamics to absorbing the cold waste products of cellular physiology. By drawing life outward, death opens life to remote systems, touching all of the resources supporting life and approaching the virtual that is recognizable but not comprehensible. As lifecycles spin, death peels off corpses into their thermodynamic sink and seeds new generations; as life dissipates energy, organisms achieve near-periodic functions in phase space, skirting the edges of chaos and gravitating toward strange attractors.

Playing the odds against life may seem like a losing proposition, but frequently the odds favor life. One might think that we are all playing the same game, but the odds favor females as opposed to males and juveniles as opposed to adults. The odds of living longer also change over a lifetime: they go up after birth, change slope after the juvenile years, plunge in adulthood (doubling every seven to eight years), and level off at a high but stable rate in old age.

Thus, in the gambling house of life, it would seem, winning requires placing one's chips, namely cells, down at every stage of life—the embryo, fetus, neonate, juvenile, and so on—while not falling victim to chance—cancer, infarction, and so on. Human beings have been moderately good winners, but organisms that add to their supply of chips in the form of additional replacements for differentiated cells raise the stakes. These organisms play for indefinite growth and negligible senescence.

Alternatively, a living thing, like a retail business, must operate at a profit just to keep ahead of creditors and its debt to inherited, accidental, and operational costs. Profits, or health, are derived from customers who also constantly reduce the stock of merchandise the way cells perform life's function and are constantly eliminated from tissues. But, while management can resupply its shelves with merchandise brought in from outside sources, the organism must resupply its tissues with cells made by the division of preexisting cells. And thus, in the balance between the cost of warehousing and having sufficient merchandise on hand, each species' cellular resupply mechanism operates at a profit or suffers the consequence.

Part II

How Death Evolves and
Where It Is Heading

Part 1 eliminated many things that death isn't while advancing the case that death is an evolving part of life. Part 2 examines mechanisms for death's evolution reduced to the cellular level of complexity, while the plot thickens around evolution and questions of consequences at higher levels of complexity.

Chapter 5 looks at the role of cells in development, maintenance, and regeneration, along side cells' role in death. Cells divide and accumulate, differentiate and migrate, but cells also contribute to health via normal turnover and die. And we die when our cellular resources are exhausted.

Several chords seem to have been struck, and chapter 6 picks up the bagpipe and self-renewing stem cell to sound out how our tendency toward longer life may be attributable to our retaining juvenile tendencies. Have we accumulated greater stores of stem cells at the expense of germ cells as we evolved, thereby obtaining superior cellular resources later into life at the expense of fecundity?

An afterword draws these chords together with a practical perspective. In brief, planning for social welfare in the future would seem inadequate if present trends in human aging continue. Indeed, future human beings might join the ranks of other animals with indefinite life spans and negligible senescence, living out their youthful potential for a thousand years or more.

Chapter 5

Putting Cells in the Picture

... for all its objectivity science, by definition, is a human construct, and offers no promise of final answers.

—Simon Conway Morris, *Life's Solution: Inevitable Humans in a Lonely Universe*

Scientists need to accept life's beauty . . . changing the focus to fit the particular needs of the particular circumstances at different times. Narratives change. In the past science has often responded to what it knew by telling stories about the world and finding the facts to confirm them. Today, the facts are telling a new story.

—Michael Boulter, *Extinction: Evolution and the End of Man*

Many pieces of death's puzzle have now fallen into place: from dissipative structures far from thermodynamic equilibrium at the edges of chaos to odds making and supply-sided economics.[1] But the puzzle is still incomplete. Since the nineteenth century, cells have been required to make sense of life. Chapter 5 puts cells into the picture of death.

CELLULAR THEORIES OF LIFE AND DEATH

In 1839, Theodor Schwann (1810–1882) presented the world with a reductionist theory of biological development. Schwann's cell theory is *not* the notion that organisms are composed (*zusammengesetze*) *of* cells, although they are, or even that differences among tissues can be attributed to substances contained in cells or produced by them. Henri Dutrochet (1776–1847), among others (Robert Hooke [1635–1703], Casper Friedrich Wolff [1733–1794], Lorenz Oken [1779–1851], Robert Brown [1773–1858], Johannes Evangelista Purkinje [1787–1869], and Felix Dujardin [1801–1862]) had already circulated those notions. By tracing tissues microscopically, from their state of complete development to their primary condition, Schwann reversed the course of development and came to the revolutionary conclusion that the growth (*Wachsthum*) of all plants and animals depended on the same elementary parts, namely cells. Schwann's cell theory is that all living things are made *by* cells.[2]

In subsequent years, the cellular theory was extended from the development of organisms to the maintenance of organisms. One can now list the number of times particular tissues are replaced in one year: lining of the small intestine, 228; lining of the stomach, 193; epidermal covering of the lips, 25; hepatocytes of the liver, 18; lining of the trachea, 8; and lining of the bladder, 6.[3] And all this replacement is perfectly normal and not a consequence of trauma.

Ultimately, biologists recognized the role of cells in death as well as life: a living thing dies when its cells no longer develop or maintain it. Death may occur at any stage of life and may have any number of underlying causes, but death is due to a deficit in cellular dynamics. A multicellular embryo dies when its cells are unable to sustain its development, and a multicellular adult, such as you or me, dies when its cells are unable to sustain its maintenance. Death due to trauma would seem an entirely different matter, but even a traumatized organism dies because its cells cannot sustain life by repairing damage to tissues or organs fast enough. Thus, the reductionist cellular theory of death, like the comparable cellular theory of life, attributes a phenomenon at the organismic level—death in this case—to phenomena at the cellular level—the absence or inaction of cells.

Can cells also explain the continuous drop-off in our expectations of life (the Gompertzian exponential)? In fact, all cells can do is divide, produce products of differentiation, migrate or remain in place, and die, but the combination of these activities over the course of time are described by a curve. Remarkably, for the years between puberty and senescence, this curve for human beings parallels the Gompertzian exponential.

This biological observation, that the empirically determined specific growth rate appears to decay proportionally with time, yields a simple model of growth, according to which the organism grows by some fundamentally exponential process, which undergoes the observed decay over its whole range. The most probable source of an exponential growth process is the self-multiplication of cells; the source of the decay, however, is difficult to identify. Cell death has been described . . . as a normal morphogenetic mechanism. . . . In addition to, or instead of, cell death, gradual prolongation of intermitotic times could produce the growth curve we observe. Loss of proliferating cells to a pool of differentiated, non-dividing cells has also been regarded as a mechanism of growth retardation.[4]

What remains, therefore, is working out the details: how the structure and activities of cells can determine the parameters of death.

THE CELL'S ROLE IN GROWTH AND DEVELOPMENT

As a rule, growing multicellular organisms, such as vertebrate embryos and fetuses, as well as regeneration blastemas, confront two problems: (1) producing large numbers of cells from small numbers and (2) employing general, homogeneous, and undifferentiated cells in the creation of increasingly specific, heterogeneous, and differentiated tissues and organs. The solution to the problem of producing large numbers of cells is for cells to specialize in symmetric divisions wherein both new cells tend to remain proliferative, thereby directing a preponderance of cells toward clonal expansion. The solution to the second problem is for the cells produced by clonal expansion to fan out across a broad spectrum of developmental potentialities while becoming committed to particular lines of determination and filling many specific, heterogeneous, and differentiated niches in tissues and organs.

MAKING AN EMBRYO

Cells make an embryo by devoting themselves to symmetric division and diverting increasingly specified portions of the population to germ layers. Even after a rodent blastocyst makes its way down the uterine tube to the uterus, it will divert only about three cells of sixty-four, or thereabouts, to forming the embryo proper.[5] These cells are part of the inner cell mass (ICM) already ensconced in the trophectoderm or embryonic portion of the future placenta (the chorion).

The premier virtue of ICM cells is their ability to produce abundant cells with the competences of embryonic germ layers ready to traverse all the developmental pathways that cross the amnion, allantois, yolk sac, epiblast and bilaminar embryonic plate. The vast range of cell- and tissue-types formed by derivatives of the ICM epitomizes *pluripotency* and is only exceeded by the *totipotency* of the intact blastocyst in utero or the combination of embryonic stem cells and a trophectodermal shell transferred to a receptive uterus.[6]

Following implantation, embryonic germ layers appear and form all the extraembryonic membranes, the fetal portion of the deciduate placenta, and the embryo—that's totipotency!

The production of embryonic germ layers commences at gastrulation; it is probably the hardest job cells ever perform and a huge boost to further development. Once established, embryonic germ layers, namely, ectoderm, mesoderm, and endoderm grow and support the development of all tissues in each embryonic rudiment: *Endoderm* and *cutaneous ectoderm* develop into a host of epithelia and their derivatives; *mesoderm* forms a broad range of tissues from epithelia to connective tissues, muscle, and, uniquely, blood; *neural ectoderm* forms a range of neurons and glial cells as well as *placodes* and the *neural crest*.

Germ layers, as such, cease to exist, however, when they take part in the morphogenesis of fetal rudiments. At that time, one speaks of a tissue as "derived" from a particular germ layer but no longer as being part of a germ layer. The organism is covered, after all, with an epidermis, not an ectoderm, and the organism is lined with mucous membranes and not an endoderm. Similarly, the internal derivatives of embryonic rudiments are no longer neural ectoderm, neural crest, and mesoderm. Embryonic germ layers, thus, are transient parts of the short-lived embryo. The cells of embryonic germ layers, therefore, are not comparable to self-renewing stem cells (see below) that remain in adult tissues and maintain them for the duration of a lifetime.

Making Tissues and Organs

Many tissues and organs are made following the delineation of self-renewing stem cells. Tissues are defined as the composite of cells and extracellular materials that are roughly (and "roughly" doesn't come close in some cases) similar in structure and function. In addition to the classics—epithelia, connective, muscle, and nerve—tissues now include blood cells (plus lymphatic cells) and germ (reproductive) cells, often with unique stem cells, basic cell types, and, sometimes, reserve cells.

Epithelia contain polarized cells broadly in contact with each other and mounted on a *basal lamella* (the epithelia's form of extracellular material). Epithelia are also characterized by cell-to-cell junctions. Connective tissue has the opposite qualities: minimum cell-to-cell contact and maximum extracellular material. Connective tissue generally "connects" blood vessels to epithelia but also comprises skeletal elements. Muscle is the contractile tissue, surrounded by its own extracellular material called a *peripheral lamella*, and neurons are conductive tissue elements supported by astrocytes and neurolemmacytes which, with their own extracellular material, surround peripheral nerves and, like oligodendrocytes in the central nervous system, envelop axons in myelin sheaths.

Blood and lymphatic cells are derived from hematopoietic stem cells (HSCs), and germ line cells are derived from primordial germ cells (PGCs). Plasma, blood's extracellular material, contains proteins secreted elsewhere (for example, albumin produced in the liver and immunoglobulins produced by sequestered lymphocytes). Extracellular material called the *zona pellucida* surrounds oocytes, and decapacitating proteins coat spermatozoa until they are capacitated in the female reproductive tract (or culture medium in the case of in vitro fertilization).

One thinks of embryonic cells as emergent and growing, but, ultimately, embryonic cells must settle down and take their place in the tissues of rudimentary organs.[7] The cell is not, after all, "born" determinate. It becomes determined. Although most research on cellular determination assumes a progressive role for control and regulative genes, a host of epigenetic controls would also seem to be at work. For example, gene silencing through DNA methylation plays a central role in modulating patterns of embryonic cell determination.[8] Indeed, delayed DNA methylation keeps embryonic cells in their compartment, while PGCs become committed through methylation. In fact, the differences between embryonic cells, PGCs, and embryonic germ cells reside at least in part in the methylation of their DNA.[9]

When does the determination of embryonic cells begin, which is to say, how far back might determined cells be traced in growing organisms? The answer is presumably different for different germ layers and the kinds of tissues formed. Endoderm, which forms epithelial tissue exclusively, would seem to become determined first, since endodermal hypoblast is the first differentiated tissue to form. Ectoderm is clearly influenced by induction from underlying tissue (endoderm), and, it would seem, does not become determined prior to receiving that influence. Ectoderm forms epithelia and nerve, primarily, but the neural crest that deepithelializes from ectoderm forms connective tissue and muscle as well. Mesoderm would seem to be determined last, if only because so much of what is formed by mesoderm takes shape processively and late in embryonic development. With the exception of nerve and the germ line, mesoderm forms representatives of each of the classic tissues, and, uniquely, forms blood and lymphatic cells. Remarkably, the germ line is segregated early but is not an indigenous part of the gonad. Germ line cells invade the mesodermally derived gonad and settle into primordial follicles or seminiferous tubules during fetal development.

Vertebrate organs acquire form from their tissues, although a dash of foreign cells, typically of neural crest origin, is virtually universal. *Indigenous organs*—conspicuously, epithelial organs and neural epithelial structures as well as large muscle masses—develop from local resources.[10] *Stratified-composite organs*, such as the intestinal and respiratory tracts, the integument, and parts of the urinary and reproductive systems, are formed when layers of tissue—epithelia, muscle, and connective tissue with their own extracellular matrices—meet, interact, and fuse. *Colonized organs*, namely the bone marrow and lymph glands, ovary and testis, are formed when a connective tissue or epithelial matrix, in the case of the thymus, is invaded and taken over (functionally dominated) by foreign tissue (hemato/lymphopoietic or germ line).

Neural crest cells are the salt and pepper of organs. These cells spread out throughout the body and differentiate into the neurons and satellite cells of ganglia, sensory cells, peripheral neurons, smooth muscle, pigment cells, neuroendocrine cells, neurolemmacytes, and connective tissue derivates of mesenchyme, conspicuously in the head where neural crest becomes mesenchyme and differentiates across a broad spectrum of cell types, including the odontoblasts of teeth.[11]

Indigenous organs are moderately indeterminate, employing induction and other forms of local and organismic interactions in the course of their development. Stratified-composite organs would seem to lean more heavily on tissue interactions, utilizing one or another tissue as a highway while getting the right tissue to the right place for induction, fusion, and differentiation.

But colonized organs are the epitome of indeterminacy. HSCs colonize several embryonic sites, beginning with a vascular endothelial/mesenchymal site and moving to the yolk sac, liver, kidney, spleen, and bone marrow,[12] and giving rise to all sorts of stage-specific blood and lymphatic cells.[13] Likewise, the primordial germ cells that colonize the primitive gonad take over the remnants of the mesonephric kidney, leaving virtually no room for the indigenous cells, although "interstitial cells" of local origin secrete hormones, epithelialized supportive cells function in germ-line maintenance, and connective tissue (stroma) provides gonadal structure and access to circulation.

Even after birth, "recolonizing cells" with embryo-like qualities can contribute to colonized organs, as demonstrated by the successful therapeutic recolonization of depleted bone marrow by cells from young as well as adult donors of bone marrow.[14] Similarly, depleted rodent testes are recolonized by presumptive spermatogonia from both pups and adults. The "homing instinct" present in recolonizing cells would seem to persist undiminished from young to adult animals. Presumably, these cells could be traced back into the fetus and embryo, although they have no "home" prior to the emergence of colonization sites or *"stem-cell niches."*

THE CELL'S ROLE IN MAINTENANCE AND REGENERATION OF ADULT TISSUES

In general, the dissipative systems we call living things hold themselves together far from equilibrium by utilizing the ultimate products of stem cells to restore cells just as fast as they are lost. Classically, adult tissues are distinguished by three types of cellular dynamics: (1) steady-state or regenerative, (2) static, and (3) expanding. *Steady-state* or *regenerative tissues* consist of a permanent subpopulation of tissue-specific *self-renewing (SR)* or adult stem cells (also known as *actual* or *functional stem cells*), and transient (impermanent) subpopulations of *transit amplifying (TA)* cells (also known as *progenitor* or *proliferative precursor cells*[15]) and a further subpopulation of differentiating (or maturing) cells. *Static* and *expanding tissues*[16] lack stem cells, sensu stricto, but may contain *stem-cell cognates* capable of performing homeostatic and regenerative functions. Static tissues do not ordinarily contain dividing cells, while expanding tissues consist entirely of differentiated cells capable of proliferation and hence maintaining normal tissue during turnover and regenerating tissue following trauma.

ADULT TISSUES

Steady-state adult tissues include epithelia of the epidermis, of the gut and respiratory tract, exocrine and endocrine glands,[17] the chondroblasts and osteoblasts of structural connective tissue, hemato/lymphopoietic tissue of blood (for example, bone marrow and lymph nodules), and the male germ line. Each of these tissues contains a small, sometimes hidden population of SR cells, typically thought of as basal cells, and much larger populations of TA cells and mature cells undergoing progressive differentiation. The consequence of SR cell loss can be dire, for instance, when the hematopoietic system is no longer provided with healthy, new TA cells.[18]

In *static tissues*—including muscle, most nerve tissue, adult chondrocytes, and, possibly, mammalian oocytes—no cell is supposed to divide, or certainly not to survive after dividing. Nevertheless, cell division is sometimes elicited in static tissues (such as the smooth muscle cells of the gravid uterus), and tissues once classified as static may turn out to be steady-state. For example, the discovery of genuine neural stem cells (preneuron/astroctyes) in the lining of the lateral ventricles of the brain has caused a revision of nerve tissue classification. Ambiguity remains, however, since the bona fide neurogenic stem cells of rat brains migrate and differentiate, while those of the human brain remain in place as glial cells.[19] Indeed, even the adult mammalian female germ line, once famous as an example of stasis, turns out to be capable of proliferation, at least in young rabbits, prosimians, and rodents.[20]

Expanding tissues, including endothelial cells, fibroblasts, osteocytes, hepatocytes, and possibly cardiac myocytes,[21] are not expanding in a literal sense—only in a potential sense. Cell division is ordinarily uncommon, especially in a mitotically quiescent stroma (connective tissue), but, whether or not they do so ordinarily, under stress cells of expanding tissues become regenerative and the population expands (liver hepatocytes and pancreatic islet cells following surgical ablation[22]). Each cell would seem capable of participating in regeneration, although most differentiated cells may ordinarily have ceased cycling.

Adult Cells

Adult Stem Cells

Adult stem cells, in the strict sense, are SR cells present in adult steady-state tissues. There, stem cells operate in normal maintenance or homeostasis, balancing cellular loss with gain, and regenerating the tissue by adjusting the balance of cellular loss and gain. Moreover, while some steady-state tissues are nearly stable throughout a lifetime, some have cyclic growth in the normal course of events. In addition to the seminiferous tubules that cycle annually in some mammals (rams[23]), hair follicles may cycle continuously between growth and stasis.[24]

Self-renewing cells typically support one or more related lineages of TA and differentiating cell. Their differentiation remains within the confines of the tissue and the SR cells are spoken of as multi- or oligo-potent. Even in the epidermis, so-called bulge stem cells give rise to the TA cells that differentiate into the squames of soft keratin (of the stratum corneum and dandruff), hard keratin (hair, hoofs, nails), and the secretory cells of sebaceous glands.[25] Similarly, in the mouse, "the 4–6 lineage ancestor stem cells thought to exist in each adult [intestinal] crypt [produce] about 300 cells . . . per day, or about 3.3×10^5 cells . . . from each crypt in the lifetime of a mouse."[26] These cells differentiate into enteroendocrinocytes, exocrinocytes, goblet cells, and intestinal absorptive cells.

Exceptionally, HSCs give rise to additional stem cells called colony-forming units. Granulocyte/macrophage-colony forming units (GM-CFUs) are the common precursor stem populations of both neutrophils and macrophages (monocytes). Macrophages, produced at the rate of 10^9 cells per day in bone marrow, leave circulation to differentiate into osteoclasts in bone, hepatic fixed macrophage in liver, microglia in the brain, macrophage-monocyte cells in epidermis and mucous membranes, and alveolar macrophages in lung. In addition, GM-CFUs may restore the steady-state in the event of trauma.

Despite this potential expansiveness, SR cells operate under very limiting constraints. Self-renewing cells seem to receive their instructions from or while occupying stem-cell niches[27] that rise and fall in the course of development. Some niches arise early enough to offer a home for migrating embryonic cells, such as PGCs and neural crest cells, while other niches (for example, bone marrow) arise later, accommodating derived "wanderers" (hemato/lymphopoietic stem cells). Stem-cell niches also seem to become inaccessible and make recolonization more difficult in adults. For example,

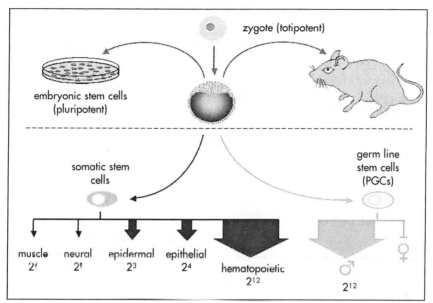

FIGURE 5.1. Stem and transit amplifying cell dynamics. Illustration kindly provided by Kyle E. Orwig, Assistant Professor, Department of Obstetrics/Gynecology and Reproductive Science, Magee Women's Hospital, University of Pittsburgh.

recolonization by identical spermatogonial stem cells is vastly more successful in the depleted testes of rat pups than rat adults.[28]

Additional constraints would seem to operate on the size of niches. Vertebrates, including most mammals, achieve and briefly maintain their maximal size some time after sexual maturity. Thereafter, internal organs are confined to the finite space available within the body, and homeostatic mechanisms come into play maintaining a dynamic equilibrium according to the requirements of the particular tissues and organs. Two caveats must be added. First, the transition between the juvenile or preadolescent and adult stage of a lifetime may not be an especially good point to look for the delineation or origin of SR cells. It is probably too late. That point of origin would probably be earlier when these cells emerge from fetal tissue.

Second, tissues in some animals do not fall within the confines of postpubescent morphological constraints. Even some vertebrates grow continuously rather than settle down to a definitive adult size. These animals presumably contain stem-cell homologues operating in the context of anatomical expansion. Vertebrates exhibiting indeterminate growth, moreover, may exhibit continuous reproduction, negligible senescence, and indefinite longevity (especially females). From fish to tetrapods, possibly including

whales among mammals, adult vertebrates exhibiting indefinite growth presumably rely on autopoietic mechanisms that govern growth and maintenance
and integrate organismic development while meeting the cellular requirements
of prolonged upkeep and repair.

The close evolutionary relationship among vertebrates exhibiting definite
and indefinite adult size suggests that their general physiology operates under
similar controls. One would expect, therefore, that organisms with definite and
indefinite growth would have different numbers of SR cells rather than different types. The control of cellular dynamics that make some organisms potentially immortal while committing other organisms to mortality might reside in
quantitative differences in SR cells.

Adult Stem-cell Cognates

While strictly speaking, stem cells are found only in steady-state tissues, *stem-
cell cognates* are found in static, expanding, and steady-state tissues. These
cognates include reserve cells capable of being coaxed into division and differentiation following traumatic tissue loss, and *cache cells* of expanding populations that act in their differentiated capacity while remaining available as a
cellular reservoir for new cells.

Reserve cells are typically nondividing and morphologically undifferentiated cells that play a role in regeneration in response to stress or trauma. The
classic example of reserve cells is the satellite cell of skeletal muscle,[29] but,
unexpectedly, astrocytes in the brain turn out to include a subpopulation of
reserve cells whose presence is only revealed following ischemia.[30] Reserve
cells are also present in steady-state tissues. For example, spermatogonia in
steady-state seminiferous tubules would seem to include a subpopulation of
quiescent reserve cells.[31] Actually, the distance between the classical stem cell
and reserve cell in steady-state tissues may not be great. For example, environmental stress in the form of nutrition restriction induces steady-state HSCs
to behave as reserve cells.[32]

Typically, cells in steady-state tissues differentiate during the G_1 phase of
the cell cycle (that is, after mitosis [M] and before undergoing DNA synthesis
[S]; see chapter 3). Reserve cells, on the other hand, come in two classes. One
class of reserve cells remains suspended in the G_0 phase (arrested after M
while not entering S) but available for recruitment into the G_1 population of
TA cells (able to enter S). The second class is suspended in G_2 (between S and
M), not regularly cycling but ready to divide immediately when activated.
Both types of reserve cells might even retain the capacity to produce new
reserve cells under stress if only for a limited number of cell divisions.[33]

Cache cells are the chief cells of an expanding tissue. They retain a cryp-
tic proliferative capacity and availability as a cellular reservoir as well as
acting in their differentiated capacity. Nowadays, cardiac myocytes seem to be

cache cells capable of getting on the cell cycle bandwagon and even acquiring markers of cardiac myoblasts and redifferentiating into cardiac myocytes, or becoming smooth muscle, or endothelium following infarction.[34] If these results are confirmed, and the cardiac myocytes available for proliferation and differentiation are not reserve cells, cardiac tissue will have to be reclassified as an expanding tissue rather than a static one.

Transit Amplifying Cells

Transit amplifying (TA) cells proliferate abundantly and vastly outnumber SR cells in steady-state adult tissues (see figure 5.1), even though they become quiescent (post-mitotic), differentiating cells and emigrate or die and are phagocytized or sloughed. Indeed, 2^{12} divisions of TA cells belonging to the erythropoietic series (proerythroblasts), the myelogenous series (myeloblasts and promyelocytes), and the lymphopoietic series (lymphoblast, prolympho-cytes, and large lymphocytes) overwhelm cell division in HSCs.[35] Similarly, 2^{12} divisions of TA spermatogonia swamp divisions in self-renewing sper-matogonial stem cells.

Two categories of TA cells are sometimes confused with stem cells. The first category, *potential stem cells* or *clonogenic cells* consists of cells capable of returning to an actual or functional stem-cell state from a TA or differenti-ated state following one or more cell divisions. Such cells are said to *dediffer-entiate* and would include lymphocytes returning to a lymphoblast state, intestinal-gland epithelial cells that re-assume stem cell properties following a disturbance, and spermatogonial TA cells that resume self-renewal prolifera-tion. The uterine mucosa of cercopithician and anthropoid primates would seem to represent a variation on this theme during its regeneration following the menstrual period. The second category, *transdifferentiating cells,* consists of cells capable of changing from one state of differentiation to another. Although well documented in jellyfish where isolated striated muscle cells can be induced to transdifferentiate into smooth muscle, and hence TA cells of nerve,[36] the closest thing to transdifferentiation in vertebrates occurs when cells fuse with differentiated cells (for example, lymphoma cells with lym-phocytes and bone marrow hematopoietic stem cells with hepatocytes).[37]

THE CELL'S ROLE IN DEATH

The notion that cells are ultimately responsible for organismic death gained legitimacy when Leonard Hayflick discovered that cells known as fibroblasts isolated from the body and grown in tissue culture gave up growing and divid-ing within a particular number of divisions (known as the "Hayflick limit").[38] These cells, said to be in a state of *replicative* (*mitotic* or *proliferative*) *senescence,* then remained dormant, but they also died after a while. Today, "[r]eplicative senescence is a block to the further replication of mitotic cells, mediated by cyclin-dependent kinase inhibitors, that leads to a viable state of indefinite cell cycle arrest."[39]

Amazingly, even bacteria[40] and budding yeast slow down and may stop dividing after a prolonged bout or particular number of divisions, and round-worms produce only a precise number of cells in their lifetime.[41] This wide-spread property of cells to run out of divisions, it would seem, could explain why so many organisms do not live forever: because their cells reach a point at which *they* do not live forever. In the case of multicellular organisms, their cells run out of the potential for cell division and cannot thereafter provide the new cells required to maintain tissues in their normal healthy state, no less repair tissues in the event of damage.

Circumstantial evidence lends support for this cellular theory of death in vertebrates: cells from older individuals divide less in tissue culture than cells from younger individuals. Moreover, cells from longer-living species (for example, tortoises) divide more than cells from shorter-living species. But the most tantalizing support comes from indirect evidence suggesting that cells have a count-down timer in their nucleus keeping track of the number of times the cells have divided. In human beings, and other mammals, this back-track-ing timer may reside in so-called telomeres, the protective knobs at the ends of nuclear chromosomes.

The Telomere/Telomerase Story

Telomeric length seems to be correlated with the number of cell divisions per-formed by cells. Furthermore, cells whose telomeres are damaged by mutation do not support the normal number of cell divisions and, indeed, organisms with these mutations experience accelerated aging.[42]

Telomeres are made of the same material as genes—DNA—but telomeres are not genes in the ordinary sense of sequences of nitrogenous bases that encode proteins or even regulatory elements of the genome. What is more, instead of the high level of specific information attributed to the sequences of nitrogenous bases, telomeres consist of the tandemly repeated, overhanging sequence TTAGGG.

Telomeres are not "junk DNA," since they seem to have a specific func-tion: binding proteins that prevent chromosomes from sticking to each other. Moreover, telomeres buffer genes against loss during normal replication as the linearized DNA of eukaryotic nuclear chromosomes shortens by about one hundred base pairs per cell division. The telomeric "timer" is thought to work by measuring the loss of telomeric repeats normally caused by successive rounds of DNA replication and blocking division after the loss of several kilobases of telomeric DNA. The theory is that shortened telomeres, or chromosomes partially damaged as a result of telomeric shortening fail to pass a "check point" in the division cycle and send the cell a "stop dividing" message.

But telomeric DNA's most peculiar eccentricity is its ability to lengthen through the action of the enzyme telomerase. This enzyme provides an RNA template and a reverse-transcriptase (an enzyme capable of RNA-dependent DNA synthesis) that lengthens telomeres thereby blocking the "stop dividing" signal and postponing replicative senescence.

Telomeric length is broadly, if loosely, correlated with a cell's potential for division. For example, the fertilized egg and embryonic stem cells of mammals are endowed with long telomeres, while cells entering replicative senescence have short telomeres. Furthermore, cells tending to divide more and those dividing indefinitely effectively express telomerase, thereby lengthening telomeres and constantly resetting the cell's division counter. For example, in rainbow trout and American lobsters, which grow throughout their lives and may age only slowly, if at all (that is, exhibit negligible senescence), telomerase occurs in cells throughout the body.[43] In mammalian adults, telomerase may provide stem cells with their protracted ability to divide and hence lengthen their life. But the telomere story may be still more complicated, since roundworms, whose adult cells do not divide, survive stress and live abnormally long lives with mutations resulting in extra long telomeres.[44]

Moreover, in some cancer cells capable of dividing indefinitely, telomerase restores or lengthens telomeres following division, and telomerase-negative cells transfected with the telomerase catalytic subunit jump into high-replicative gear after expressing telomerase and elongating telomeres.[45] Finally, the capstone of the telomere story is the discovery that otherwise normal, transgenic cells (having received artificial, foreign genes through genetic engineering) divide without limit when over-expressing telomerase and lengthening their telomeres.[46]

WHAT'S WRONG WITH THE TELOMERE/TELOMERASE STORY?

The telomere/telomerase story goes pretty far toward a cellular mechanism of life and death, but it does not go far enough. First of all, the number of cell divisions may not be a good measure of a cell's longevity in the organism. Cell division is widespread in the body, of course, notably in the TA precursors of differentiated cells,[47] but cells do not normally divide in static tissue, such as nerve and muscle, in the adult body, although they may renew their own parts through other mechanisms.

Furthermore, the correlation of telomeric length with the number of times cells move through their division cycle is not entirely straightforward. Indeed, mice, with far fewer cell cycles in their lifetime than human beings, have telomeres three times longer than those of human beings. Moreover, genetically engineered mice lacking telomerase and experiencing telomeric shortening do not exhibit generalized premature aging (such as impaired cardiovascular system, blood glucose tolerance, liver, kidney and brain structure, or demonstrable osteoporosis, arteriosclerosis, or cataract formation), although after a few generations the affected mice exhibit lowered fitness and well-being leading to shortened life span (accelerated graying, skin lesions, impaired wound healing, blood-cell kinetics, and increased numbers of spontaneous malignancies).[48] What is more, some mutant genes would seem to curtail the potential of cells to divide a normal number of times without affecting telomeric length, and some immortal cancer cells are associated with shortened telomeres and other chromosomal aberrations.

In the final analysis, as the science writer Stephen Hall puts it, "aging . . . was not simply a matter of telomeres, and perhaps not even principally a matter of telomere biology. Genes, cellular metabolism, caloric intake, DNA damage—they all seemed to play important roles, and there was no scientific agreement on what *caused* aging."[49] Hall concludes by quoting Lenny Guarente, the MIT specialist in yeast/roundworm aging, saying, "So, if one thinks dispassionately about it, there's no real reason to think that telomeres are doing anything."[50]

Indeed, the telomere/telomerase story may conceal a chicken and the egg conundrum: which comes first, telomeric length or telomerase activity? Where telomeres are shortened, their length may play a role in limiting the cell's life span, but the active principle or culprit of the story may be the enzyme, telomerase, that lengthens telomeres, and not the number of cell divisions that shortens telomeres.

In any event, it would seem telomeres and telomerase are not the whole story, even if they are part of it. Recent evidence suggests that additional complications, if not uncertainties, arise from mitochondria.[51] In particular, mutant mitochondrial polymerase promoting the accumulation of errors in mitochondrial DNA (mtDNA) accelerates cellular aging and cell death.

The question is, what else is involved in cellular life–extension/shortening? Answers may be sought in a variety of areas: examining animals with negligible senescence and an indeterminate life span; discovering how cancer cells can divide indefinitely; exploring how eggs and spermatozoa reset their counter of cell divisions and form a fertilized egg capable of initiating divisions with a fresh slate; and studying how stem cells acquire a greater potential for division than ordinary body cells.

THE CELL'S POTENTIAL ROLE IN REGENERATION THERAPY

Stem cells obviously fill a large gap in the enigma of organismic life, its development, maintenance, aging, dying, and death. Normally, throughout a lifetime, cells are the only source of cells that resupply the organism with requisite replacements cells. We live as long as stem cells and their cognates are able to sustain life by providing TA cells and differentiated cells during normal turnover while also meeting the challenges of healing or regeneration following trauma. On the other hand, apoptosis and loss of stem cells would seem to play a role in organismal aging.[52] When compensatory proliferation is too low, the stem cell pool runs empty, normal function in tissues comes to a grinding halt, and the organism becomes a corpse.[53]

The question is, can the artificial introduction of stem cells prevent the dire consequences of stem-cell loss? Unfortunately, this question is more difficult to answer than one would hope. The difficulties, if not the answer, can be sorted out through a historic reconstruction and analysis of the concept of stemness.

A SHORT HISTORY OF STEM CELLS

The term "stem" has several roots in biology. Conspicuously, in botany, stems are aerial axes of plants generally produced by the *meristem* or the growing part of a stem where small, dividing cells give rise to initiating cells and hence derivative cells that differentiate into all plant tissues. The biological stem can also be the stock or main ancestral line that gives rise to the branches of a family or a fundamental or primitive group from which other members of a clade may have evolved.

Surprisingly, "stem cell" appears only twice in E. B. Wilson's 1896 edition of *The Cell* (on pages 111 and 112), in both instances referring to the primordial germ cell in *Ascaris* and *Cyclops* that alone retains intact chromosomes following cleavage of the fertilized egg.[54] In the monumental 1925 edition, Wilson adds several dipterans and higher invertebrates to his list of organisms with primordial germ cells but retains "stem cells" exclusively for cells that give rise to oogonia or spermatogonia.[55]

By the 1920s, however, stem cells were discussed as the source of particular blood cells (erythroblasts) and the concept of stemness was associated with the sources of differentiated cells from clonal lineages or colonies produced from a single cell. The issue of stem-cell potency arose regarding the possibility that "a stem cell from the bone-marrow is multipotent and, depending on the particular internal environmental niche in which it lodges, will develop into erythropoietic, granulopoietic or lymphopoietic cell lines."[56]

Following World War II and the tragic beginnings of the atomic age, radioactive markers were employed to trace cells through the *mitotic cycle* with its distinct phase of DNA synthesis (S) dividing interphase.[57] Studies on cell proliferation, the clonal origins of tissues, and on cellular dynamics flourished,[58] and, in their wake, stem cells emerged as "cell types capable of extensive self-maintenance (self renewal) *in spite of* physiological or accidental removal or loss of cells from the population."[59]

Stem cells were still broadly thought of as unipotent, giving rise to one type of differentiated cell, or multipotent, giving rise to a few types of differentiated cells associated with a tissue, but they were not thought of as pluripotent and giving rise to virtually all differentiated cells. Indeed, the stem cell might have solidified at this time around *oligopotency* (a few potencies for differentiation) and homeostasis (tissue maintenance and regulation), but it was not to be. As one modern stem-cell theorist has insisted, "[c]ells that are *unipotent,* though sometimes referred to as stem cells, should not be so described even if they retain some capacity for self-renewal."[60]

Stemness was shaken to its roots in the 1970s by the advent of monoclonal antibodies and vast improvements in microscopy and, soon, imaging utilizing fluorescence and digitalization. The new techniques set off a spate of experiments attempting to identify unique cell types, trace their lineage, and follow their movement. Putative stem cells were quickly traced back to embryonic cells, and the idea that stem cells took their origins in embryos revived interest in the notion that stem cells were *pluripotent* (having the ability to differentiate across tissue-specific lines and into cells of all three germ layers).

The term "stem cell" quickly infected the language of developmental biology and spread to embryonic germ layers. Embryos consisting of a few cells, such as those of ascidians and Spiralians (conspicuously, *C. elegans*), were said to form their embryonic germ layers from "stem-like" cells or cells with asymmetrical divisions, while "blast-like" cells with "symmetrical divisions" (also called "proliferative divisions") were said to generate lines of cells committed to specific paths of differentiation. The latter included "founder" cells (also called "embryonic blast" cells and "stem" cells) in *C. elegans,* "teloblasts" in clitellates, annelids and mollusks, and "set-aside" cells in embryos of marine invertebrates, larvae, nymphs and instars of arthropods,[61] where massive amounts of cell death accompany the development of adult organs at critical stages, molts or cataclysmic metamorphosis. Massive cellular turnover at this extreme is hardly reminiscent of the steady-state kinetics of traditional stem cells, and Donald Williamson breaks with tradition to attribute metamorphic events in larvae to unconventional evolution.[62]

"Stem cell" was also invoked to name the source of embryonic germ layers in echinoderms, amphibians and fish.[63] Only mammalian embryos resisted the trend. Monozygotic twinning suggested a symmetry or equality among early embryonic cells as opposed to the asymmetry implied by stem-cell division. The notion of stem cells infiltrated mammalian embryology tangentially, however, when rodents were found to form their inner cell mass (ICM) by horizontal (periclinal or paratangential) cell division early in development.[64] Moreover, asymmetry, if not genuine stemness, seemed to be in play when the ICM gave rise to the epiblast and it, in turn, gave rise to the hypoblast and the bilaminar embryonal plate.

The derivation of each of these rudiments from a small number of "founder" cells is *not* reminiscent of stem-cell behavior, especially since these populations are *not* self-renewing. They are transient and give rise to other embryonic rudiments. Nevertheless, today the title of "stem cell" is generally conceded to cells produced by the earliest divisions of blastomeres. Moreover, cells harvested from ICMs and epiblasts are known as *embryonic stem (ES) cells* when raised in tissue culture (see below).

The notion of embryonic pluripotency also infected the traditional notion of stem cells. Traditionally, multipotency was acknowledged in hematopoiesis, especially following discoveries suggesting that the source of the T and B types of lymphocytes as well as red blood cells, granulocytes, monocytes, and their representatives in chronic myeloid leukemia (for example, the Philadelphia chromosome positive stem cell clone) and other tumors was the same primordial HSC. Tracing this HSC back from bone marrow to liver, spleen, and yolk sac broadened the notion of multipotency further but did not stretch it to pluripotency. Instead, research on stem cells was refocused on the narrowing of potency. In particular, research turned to the control of differentiation, whether through a cell's history or its behavior, induction, circulating factors (for example, erythropoietin), transducing pathways, microenvironments, physical conditions, and organization.

Ultimately, pluripotency was placed on the agenda of stem-cell research by the success of nuclear transplant experiments (or "cloning") to alter cellular potency. Attributing pluripotency to stem cells recast them in the role of jack of all trades, and stem cells in adults were portrayed as if they were an atavism or embryonic leftover. Even the long-time stem-cell biologist Irving Weissman conceded that "[i]t is reasonable to propose that most, if not all tissue and organ systems are based on a stem and progenitor model during organogenesis."[65]

The premier evidence on behalf of pluripotential stem cells came from rodents, namely, that "cells derived from BM [bone marrow] can give rise to cells typical of other tissues . . . such as muscle, brain, heart, and liver . . . [presumably following] a multistep process entailing migration, conversion to a new phenotype, and expression of functions characteristic of the tissue in which they now reside."[66] The next step would seem to be a "proof of principle" demonstration in human beings.

Thus, stem cells and the concept of stemness came to its present muddle. Stem cells were thought of loosely as the wellspring of tissues, but whether stem cells were oligopotent or pluripotent was uncertain. Were they morphological entities or merely biological functions with no discrete cellular identity? Were stem cells members of particular tissues or were they generic cellular sources for many tissues? Were adult stem cells limited to roles in tissue maintenance and repair, or were they adult equivalents of embryonic cells with virtually unlimited roles to play in regeneration?

The different kinds of stem cells implied by these questions are not merely of interest to biologists. Stem cells have become an issue weighing heavily on the mind of bioethicists and the conscience of citizens concerned with embryonic cell research. If the adult can serve as a source for pluripotent stem cells, then there would be no need to harvest pluripotent stem cells from embryos. The various forms of regenerative therapy currently crying out for pluripotent stem cells could be satisfied without destroying embryos. If, however, pluripotent stem cells are exclusively embryonic in origin, and the adult stem cells are already committed to narrow pathways of differentiation, then one may face a moral dilemma regarding prospects for the therapeutic use of embryonic stem cells.

ON THE CONCEPT OF STEMNESS

Notions of stemness seem to come out of two different biological traditions, namely traditions of determinate and indeterminate morphology. Neither tradition is especially accepting of the other, and researchers in regenerative medicine are torn between them while trying to advance the application of stem-cell theory to therapy.

Determinate Stem Cells

Difficulties defining stem cells were apparent as early as 1979, when cell biologist Christopher Potten pointed out that "stem cells cannot be reliably morphologically identified and their study is restricted to various functional tests."[67] In 1990, cytologist Markus Loeffler joined Potten in placing stem cells at the center of a biological uncertainty principle: "Here, we find ourselves in a circular situation: in order to answer the question whether a cell is a stem cell we have to alter its circumstances and in doing so inevitably lose the original cell."[68]

Today, some stem cells, such as the spermatogonial stem cell (in, for example, *Drosophila*) are defined morphologically, and morphology continues to provide tantalizing hints toward the identity of other stem cells. But the hope, if not expectation, of identifying stem cells specifically with the aid of antigens and fluorescent markers has proven elusive. Markers for alkaline phosphatase, the transcription factor Oct-4, stage-specific embryonic antigens

(SSEA-3 and SSEA-4, and TRA-1–60 and TRA-1–81), and other cell surface antigens (whether present or absent) were successfully employed in stem-cell enrichment protocols utilizing the fluorescence-activated cell sorter (FACS).[69] But massive efforts, involving hundreds of markers, failed to identify antigens uniquely expressed in stem cells as opposed to combinations—molecular signatures—of antigens enriched in stem-cell populations. Indeed, "there are only *six* genes [antigens] shared between the sets identified by [two groups of researchers, while a] . . . third group w[as] able to identify only *one* gene that appeared on all three lists of genes for 'stemness'!"[70]

So-called side population protocols utilizing the exclusion of the DNA binding dye, Hoechst 33342, and FACS have also been useful for enriching inoculants with putative stem cells (for example, cells low or negative for CD34). Furthermore, slowly dividing, so-called label-retaining cells (LRCs) that retain a nucleotide analog (bromodeoxyuridine or tritiated [³H] thymidine) might very well turn out to be stem cells.[71] Undoubtedly, "specific cell surface markers would be useful to identify stem cells definitively, compare them across tissues, and distinguish them from other cells, . . . [but] currently only enrichment, rather than purification protocols exist for most tissues."[72]

Nevertheless, the determinate stem cell may yet be recognized by its competence for self-renewal and leave-taking differentiation. Known as *asymmetric division*, a dividing stem cell produces an SR cell that remains in the stem-cell population and a TA cell that moves into the population committed to differentiation.

The ability of stem cells to undergo asymmetric division is not found in cells of any other cell type and uniquely allows stem cells to maintain their own population while refreshing tissues and organs with a stream of replacement cells. Moreover, the onerous task of maintaining the integrity of various tissues and organs, and hence sustaining the organism, is played by stem cells for the duration of a lifetime. Indeed, "[f]or readers who are not stem cell biologists, it is pertinent [to point out] that stem cells [retain] . . . the continued capacity to proliferate during adult life (unlike mammalian primordial germ cells . . .)."[73]

A cell's fate following asymmetric division may be either loosely governed, which is to say, decided stochastically, or tightly governed as a function of physical attributes of the two new cells. A cell's position, self-feedback, autocrine influences, or more remote paracrine influences such as gradients[74] may decide which of two new cells remains in the stem-cell population and which differentiates. Conditions in a stem-cell niche[75] or adhesion to "anchor" or focal sites[76] may influence the cell's decision to divide in the first place, as well as which new cell remains in the stem-cell population and which is committed to differentiation. The new stem cell may even be the cell retaining template strands of DNA, while the cell destined to enter the TA population may be the cell acquiring newly replicated strands.[77]

Stem cells defined by asymmetric division are determinate in the mathematical sense of having exact and definite limits, but confusion abounds surrounding stem cells' place in the scheme of biological differentiation. Stem cells are sometimes said to be undifferentiated, although they can give rise to differentiated cells. For example, stem cells in steady-state tissues such as the mammalian epidermis are sometimes considered undifferentiated cells of the stratum basale. On the other hand, the same stem cells are recognized as both proliferative and differentiated basal keratinocytes containing low formula-weight varieties of keratin. Indeed, in some steady-state tissues, stem cells are conspicuously differentiated. For example, nonciliated bronchiolar epithelial cells and type II pneumocytes synthesize and release differentiated products (anti-inflammatories and surfactant, respectively) while, at the same time, supplying the proliferative precursor cells that become terminally differentiated when they cease dividing (ciliated pseudostratified columnar epithelium and type I pneumocytes, respectively). These "differentiated" stem cells can also serve as the source of epithelium during the remodeling of the pulmonary tree following trauma.

Indeterminate Stem Cells

The allure of stem cells for many contemporary researchers is not the ability of stem cells to make binary choices between renewal (that is, returning a cell to the stem-cell population) and differentiation (turning on the predetermined program for leaving the stem-cell population). The allure is the possibility that stem cells can undergo "fate switching," that is, exhibit pluripotentiality or indeterminacy[78] (the phenomenon of naive cells differentiating in any of several directions) or plasticity (the ability of previously committed cells to differentiate along a new pathway).

Stem cells might acquire pluripotentiality as a consequence of dividing in series and making successive binary decisions toward new pathways of differentiation, but division as such would only permit and not direct fate switching.[79] Were it possible to reprogram putative stem cells and direct them along desired paths of differentiation, it might be possible to provide cells capable of restoring or regenerating adult tissues or organs.[80] Pluripotential cells thus created would be enormously valuable for therapeutic purposes, and advancing stem-cell research along these lines would seem important for human health care.

But problems abound. In the first place, indeterminate, pluripotential cells are typically obtained in small numbers and are generally raised to usefully large numbers of cells in tissue culture (aka in vitro). This practice began with embryonic carcinoma (EC) cells and EC cell (ECC) lines originating in tumors of gonadal origin and maintained through passage in vivo (via inoculation of animals) and in vitro.[81] Stemness was attributed to these cells when some were found to differentiate in teratocarcinomas in vivo, as various tissues in vitro under particular circumstances, and even take part in embryo formation following introduction into blastocysts. Other ECC lines ran the gamut between mortal blast cells that differentiated and ceased dividing and immortal transformed cells that divided, did not differentiate, and proved cancer-forming upon reintroduction to normal animals.

Cells obtained from the ICM and epiblast of blastocysts and maintained in tissue culture have also proved to be pluripotent and indeterminate upon reintroduction to blastocysts and following "tweaking" in vitro (becoming trophectoderm capable of synthesizing human chorionic gonadotropin; embryoid bodies and cells expressing markers for neural precursor cells, and rhythmically contracting cardiac muscle).[82] These cells are better known as *embryonic stem (ES) cells*. They maintain a normal karyotype, remain proliferative, acquire a rounded (rodent) or flattened (human) appearance, and form spherical colonies (rodent) or fascicles (human). Similarly, tissue culture cells originating from rudimentary gonadal ridges have become known as *embryonic germ (EG)* or *germ stem (GS)* cells if they resemble ES cells.

Problems

Asymmetric divisions, self-renewal, and differentiation, on the one hand, and pluripotentiality and fate switching, on the other, cannot be present simultaneously in the same cell. Moreover, whether embryonic or adult, the determinate stem cell is not a naïve, undifferentiated cell on its own. It is a potentially proliferative member of a prospective or discrete tissue, and it is already committed to the differential properties of that tissue. Fate switching and pluripotentiality, whether in vitro or in vivo, would require undoing the commitment to differentiate along given lines already built into the determinate stem cell. Researchers must decide what type of properties a stem cell has if they are to avoid foisting one set of virtues upon cells with an entirely different and incompatible set of virtues.

Are ES cells normally present in adults? Even Irving Weissman, the champion of the determinate stem cell, has conceded "that stem cells are retained throughout life to participate in regeneration and repair."[83]

But do adult tissues normally contain ES cells? Maybe not, since presumptive HSCs do not exhibit pluripotency following transplantation to blastocysts, although the progeny of the HSCs may exhibit some reprogramming of gene expression.[84] If cells operating as adult stem cells arise late in development and act principally in tissue renewal, it would not seem possible for them to be "embryonic."

Redefining Stem Cells Operationally

How does one decide when a stem cell is a stem cell? The difficulty might never have arisen were embryos distinguished from adults and a transition between development and homeostasis recognized explicitly. One expedient would be to identify self-renewing (SR) cells of steady-state tissues and adding "adult" (A) to the names of their cognates (adult reserve [AR] and adult cache [AC] cells), thereby recognizing determinate qualities and the participation of cells in renewal and regeneration. The question of stem cells' indeterminate qualities, their pluripotentiality or plasticity—differentiating into a variety of cell types upon reintroduction into normal or cellularly depleted adults—would remain open and left to empirical testing on a case-by-case basis.[85]

Whether in the embryo, juvenile, or young, old or senescent adult, actual stem cells would comprise small subpopulations of cells with low rates of proliferation, and only one cell on average for every two cells produced by division would remain in the stem-cell subpopulation. HSCs in bone marrow, for example, are so rare that they may escape detection entirely in tissue culture, and, even at that, estimates of their number by experimental reconstruction with purified or retrovirally marked cells seems to have exaggerated HSC numbers by as much as twentyfold. Indeed, SR cells divide far less often than generally assumed, and, compared to TA cells, SR cells maximize their G_0 reprieve.

Consequently, SR cells would seem to have ample opportunity to correct errors of replication rather than contribute damaged DNA to future cell populations, and stem cells would seem to be in no danger of suffering from excessive telomere shortening or replicative senescence (that is, exceeding their Hayflick limit). On the other hand, the terminal differentiation of TA cell progeny would seem perfectly compatible with cells entering mitotic senescence without thereby incurring any penalty, especially if the post-mitotic cell is ultimately shed or destroyed.

Normal turnover of SR cells—such as those of epithelia, spermatogonia, hemato/lymphopoietic tissue, and (surprisingly) neurogenic stem cells—maintains the cellular balance of steady-state tissues remarkably well, replacing effete cells and renewing the stem-cell population, thereby restoring vigor to tissues and organs. Indeed, transplanted in series, SR cells may be sustained throughout several lifetimes despite telomeric shortening![86]

Stem cells exhibiting "immortal self-renewing properties"[87] offer an infinite ability to maintain and restore tissues and organs. Thus, stem cells have acquired a reputation as the fountains of youth, or the antithesis of aging. What is more, in instances of regeneration, stem cells and their cognates play the role of redundant element or backup.[88] For example, the satellite cells present in skeletal muscle and possibly myocytes present in cardiac muscle may be mobilized by trauma even if they do not ordinarily undergo cellular turnover.

In effect, stem cells and their cognates seem to be part of homeostatic self-maintenance, while their homologues govern indefinite growth. These cells develop from germinal populations in the embryo and fetus but are otherwise independent of developmental systems. Pluripotentiality is not a typical feature of stem cells and their cognates. Rather, self-renewal and differentiation along predetermined pathways are the cells' virtues in organismic maintenance and regeneration.

Embryonic Stem Cells

Pluripotentiality is the chief point of departure for much of the research in stem cells. Indeed, pluripotentiality is supposed to be a quality shared by ICM cells and blastocyst-derived "embryonic" cells in tissue culture called *embryonic stem (ES) cells.*[89]

The equation of ICM cells with ES cells implies that embryonic cells remain in germ layers and even in the adult as SR cells—hidden but virtually unspoiled, untouched by adult life, and ready to give rise to tissues whenever and wherever the need arises. Indeed, ES cells are often called *primitive stem cells*[90] in the contemporary literature, thereby removing any trace of their derivation and semantically suggesting that the same cells perform functions in both embryonic developmental and adult maintenance.

This inference violates several valid principles of biology, including the tenets that organs operate under physiological restraints and evolve under structural constraints that do not accommodate excess. Furthermore, an organism's history unfolds processively and does not stop in medias res. Indeed, one is hard-pressed to find any example of cell lineages that are unchanged between embryo and adult. Even the red blood cells produced from HSCs in different parts of the embryo, fetus, and adult synthesize different hemoglobins, and this epitome of stemness undergoes age-related change.

In practice, ES cells are tissue-culture cells derived from the twenty to thirty cells of the ICM of the late, pre-implantation blastocyst, placed in culture and exposed to feeder-cell layers or a battery of specific factors, such as the cytokine leukemia inhibitory factor (LIF) and basic fibroblast growth factor (bFGF). Likewise, EG cells are tissue-culture cells, presumably derived from primordial germ cells (PGCs) in gonadal rudiments placed in culture and exposed to a battery of cytokines and growth factors.[91] Both ES and EG cell lines are considered pluripotent when they produce tissues of all three embryonic germ layers following transplantation into blastocysts and when tweaked into differentiation in vitro by the withdrawal of some factors (LIF) and the addition of other factors. For example, some putative ES cells develop cardiac myocyte markers when treated with retinoic acid, ectodermal growth factor, hepatocytes growth factor, bone morphogenetic protein-4, and bFGF.

Of course, the technique of tissue culture has its own rationales, and practitioners have their own arcana and argot. Thus, ES and EG cells are also defined by the presence and absence of a variety of cell markers—their *transcriptome*—as general as alkaline phosphatase and the expression of the transcription factor Oct-4 and as specific as a complex cocktail of fluorescent markers used to detect and sort cells via FACS.

Undoubtedly, cell markers offer an excellent opportunity for tracking novel gene activity and for tracing the transcriptome through the serial analysis of gene expression (SAGE). Remarkable progress has been made identifying receptors for cytokines and growth factors and hence members of upstream and downstream transduction pathways active in the course of differentiation. But even the staunchest defenders of the method will admit that a list of cell markers is not an adequate criterion of ES and EG cells or their differentiated progeny.[92]

One must also bear in mind that ES cells are first cousins of EC cells that become metastatic, invasive, and destructive cancer cells in a dose-dependent way upon introduction to blastocysts, neonates, and adults.[93] What is more, stem-cell niches normally occupied by adult stem cells may not be readily accessible to reintroduced cells, for example, where these sites are separated from circulation by extracellular material. While reintroduced cells may very well know their "home," they may not be able to reach it. Instead of differentiating under local control, in harmony with the microenvironment, reintroduced cells may simply die at ectopic sites, grow harmlessly but to no purpose, or undergo malignant transformation and metastasize. The stakes are high, but practice must not run ahead blindly: stem-cell therapy still faces a formidable cancer-cell barrier.

On the other hand, tissue culture will undoubtedly be useful in "ex vivo" genomic research, in the expansion of autologous cells, the augmentation of desirable cells (for example, cancer-depleted marrow or cancer-deficient mobilized peripheral blood), and possibly transdifferentiation (skeletal to cardiac muscle) and restoration therapy.[94] And, if ES cells can, indeed, be moved to organisms from tissue culture, can travel to, arrive at, and lodge in appropriate stem-cell niches in an undifferentiated state and generate cells of one or more appropriately differentiated types, then these ES cells will have enormous impact on how anti-aging medicine is practiced in the future—to say nothing of therapy for chronic, degenerative disease and trauma.

IN SUM

Life, it would seem, boils down to utilizing one's cellular resources (merchandise) wisely (or placing one's chips on the best bets available), but wisdom and utility are relative and may change at different stages of a life.

Thus the developing embryo, fetus, neonate, juvenile, and young adult require cells predominantly as growth and accrual material, while adults require cells predominantly as replacement material (in exchange for effete, differentiated somatic cells), while, at the same time, avoiding cancer, infarction (especially in cardiac tissue), and other forms of obstruction to normal function. In adults, differentiation and cell death create the constant demand for cells met by TA cells arising from SR cells. In multicellular organisms, such as us, death threatens when cellular supplies are inadequate or following their exhaustion.

Why does the organism run out of cells? One possibility is that cells enter a state of proliferative senescence after reaching a replicative threshold, or Hayflick limit. This possibility is compatible with the further possibility that telomeres, or caps at the ends of chromosomes, shorten and act as countdown timers. But telomeres are also lengthened by the action of telomerase, and the control of telomerase expression has yet to be explained, particularly in stem cells. The ubiquitous distribution of telomerase in negligibly senescing organisms suggests that the regulation of stem-cell resources is subtler than the operation of a countdown timer.

Adult SR cells were discovered through the analysis of cellular dynamics in adult tissue. In steady-state adult tissues, SR cells produce both new stem cells and TA cells that go on to proliferate and create a population of differentiating or maturing cells. In static adult tissues, cells do not ordinarily divide, but a reserve cell population may retain the capacity for a limited number of divisions in the event of traumatic tissue loss. In expanding cell populations, cache cells retain the capacity to divide and can also exercise a capacity for proliferation in the wake of trauma.

On average, SR cell division is asymmetric, one cell returning to the SR cell population, the other cell moving into the TA cell population. The control of this asymmetry is obscure, but its dynamics are presumably altered during regeneration. In effect, SR cells and their cognates, AR and AC cells, are parts of homeostatic mechanisms, while homologues govern indefinite growth. Proliferation and differentiation along a few predetermined pathways encompass the TA cells' roles in organismic maintenance and regeneration.

In practice, adults depend for survival on SR cells and their cognates in steady-state, static, and expanding tissues. TA cells may divide copiously, but they are not self-renewing, and their transplantation to a damaged or depleted tissue may therefore provide immediate but only temporary relief, on the style of a blood transfusion. While bone marrow-derived hematopoietic stem cells find their way to niches and repopulate damaged or depleted hemato- and lymphopoietic sites, restoring them and the organism to good health, the ability of other SR, AR, and AC cells to "home" successfully awaits "proof of principle." Stem-cell niches may also deteriorate in older organisms.

The ES and EG cells derived from embryos or fetal tissue and raised in tissue culture would seem to have a broad capacity to differentiate into cells of many kinds—they are pluripotent. This capacity is demonstrated when these cells are introduced into blastocysts or "tweaked" in tissue culture. The therapeutic use of such cells is the goal of much research on stem cells, but the goal is far off.

Chapter 6

Neoteny and Longevity

The young of the human race show some anthropoid features, whereas the young of the chimpanzee approach more nearly to the human than the adult chimpanzee does. That seems to show that . . . our ancestors were more Simian than we are.

—Agatha Christie, *The Man in the Brown Suit*

Obviously, a transformation or metamorphosis is necessary in order that the adult organism may function (except that the neoteinic type of organism functions and grows as a larva until it attains maturity).

—N. J. Berrill, *Growth, Development, and Pattern*

The bagpipe model of life extension illustrated in chapter 2 suggests that we are living longer because our juvenile stage of development is percolating into our adult stages. Chapter 6 examines this suggestion, beginning with a discussion of the phenomenon of juvenilization known in the evolutionary literature as *neoteny*, from the Greek meaning stretching ("extending" or "holding onto") the new or youthful.[1] The chapter goes on to suggest that increased numbers of self-renewing (SR) stemcells acquired during development might slow down the rate of aging in adults.[2] The additional SR cells might be gleaned from stocks of primordial germ cells (PGCs) that would otherwise have become germ cells. Were that the case, the current trend toward reduced fecundity may be linked to the current trend toward increased longevity.

THE TIME IS OUT OF JOINT

Neoteny belongs to the class of evolutionary mechanisms known as hetero-chronies or age-related deviations of development.[3] Heterochrony is diagnosed when deviations in timing and/or rate of development lead to asynchronies among processes or to disparities in morphology, for example, when parasites exhibit hypersexual development and morphological reduction.[4]

Caleb Finch followed the path of heterochrony to aging.[5] Taking his lead from Gavin de Beer and Stephen Jay Gould, Finch called juvenilization "pae-dogenesis" and attributed it to either of two evolutionary processes: progene-sis, or paedomorphosis, and neoteny, or fetalization.[6] Progenesis and neoteny may be thought of as running in opposite directions—one accelerating, one slowing, and one affecting larvae or juveniles, one affecting adults. In proge-nesis, sexual maturity is pushed back into the juvenile stage, whereas in neoteny, development slows and sexually mature adults retain juvenile mor-phology. For example, aphagous dipterans, aphids, and mayflies that form oocytes before hatching are progenic, as is the tiny, pedomorphic vertebrate, the infantfish, *Schindleria brevipinguis,* whose lifetime is over at two months.[7]

In contrast, Finch defines neoteny as "sexual maturation at the usual age, but with retarded development of the other somatic tissues."[8] The facultative neotenic Mexican axolotl, *Ambystoma mexicanum,* and the tiger salamander, *Ambystoma tigrinum,* which are sexually mature but morphologically larval, are the classic examples of neoteny. Adult development is suppressed by com-paratively low levels of thyroid hormone, even though thyroid hormone recep-tors are present and capable of binding exogenous hormone.[9] Neoteny is epitomized by asynchronous, slow development of larval characteristics cou-pled with the maturation of gonads. While becoming sexual, the plastic juve-nile morphology remains dominant. That is, individuals mature sexually while retaining the characteristics of youthfulness.

Historically, Albert von Kölliker is credited with proposing that larvae might acquire sexuality, while Alexandr Onufrievich Kovalevskii (Alexander Kowalevsky) first suggested "that the larval ascidian might be the actual ancestor of the vertebrates."[10] Walter Garstang then completed the loop by suggesting that the development of sexual maturity in an overgrown, swim-ming ascidian larva resulted in the loss of the original adult ascidian.[11] Con-trary to Ernst Haeckel's notion of recapitulation that would have locked developmental stages in an irreversible sequence, Garstang's "neoteny" uncoupled adult morphology from sexual maturity and allowed the latter to work with larval morphology.

Neoteny has long figured into many evolutionary schemes. The cele-brated zoologist Libbie Hyman, among others, suggested that the bilateria, which is to say most animals, originated via neoteny from larval radiates resembling cnidarian's planula larvae.[12] Indeed, fish (ice goby), amphibians (the obligate neotenic mud puppy *Necturus maculosus* and *Proteus anguinus*), birds (the flightless ratites), and altricial mammals, notably ourselves, seem to have taken the route of neoteny and adopted new evolutionary directions out of old evolutionary patterns.

Human beings seem to be experiencing evolution both by progenic (selected, accelerated growth), and neotenous (generalized slow and prolonged development) mechanisms. Progenesis is also suggested by the "[a]cceleration of the rates of maturation during the past 150 years [which] is well documented in developed countries for age at menarche, age at peak height velocity, and age at cessation of growth in stature."[13] Notwithstanding problems of collecting data via surveys, during the period of 1960 to 1970, "at least for some women—the decline [in age at menarche] would be approximately 1 year over a generation."[14] Age at menarche is, however, notoriously sensitive to a variety of genetic and environmental factors, including years of education, family income, body weight (ponderal index), and skinfold thickness.

On the other hand, neoteny would seem to have taken place inasmuch as "[h]uman beings reach puberty at an age (12 14 years) that is [relatively] 75-fold later than in mice."[15] Moreover, the gerontologists Tom Perls and David Snowdon have shown that women reaching menopause later in life (who have grown old more slowly than other women) tend to be longer-lived, presumably due to anti-aging effects of higher levels of female sex hormones.[16]

Neoteny is also invoked to explain our growth and development compared to other primates. "The baboon mortality rate doubles every four years compared to seven to eight years for humans. . . . [Thus,] [h]umans . . . age differently and more slowly than baboons."[17] According to the pioneering primatologist Sherwood Washburn, "[t]here is strong direct evidence for the slowing of development"[18] in the course of human evolution, which is to say, "[w]hat characterizes modern humans as unique is a prolongation of the postnatal growth period."[19] Indeed, "[t]he ages derived for *Australopithecus, Paranthropus* and early *Homo* described biological equivalence to modern man at roughly two-thirds the chronological age, demonstrating that they had growth periods similar to the modern great apes."[20] "At the end of growth, the adult skull in humans reaches an allometric shape (size-related shape) which is equivalent to that of juvenile chimpanzees with no permanent teeth."[21]

Slowing extends to neurobiological features of the brain and the acquisition of behavior.[22] Indeed, adult language may very well be an extension of a juvenile capacity for vocal communication.[23] Moreover, one is hardly surprised when the biographers of the French supercentenarian Jeanne Calment describe her as "someone who remains very young in spirit and tastes, a kind of kid, almost childlike at times."[24] Jeanne Calment's juvenilization would certainly be the tip of the proverbial iceberg if neoteny is pushing juvenile well-being upon adults.

Neoteny is clearly indicated by our leaning toward altricial as opposed to precocial development. Altricial development is the condition of some birds and mammals that are helpless at hatching or birth and dependent on a parent (or parents) for food and other resources. Precocial development, in contrast, is the condition of animals (including other birds and mammals) that are morphologically if not behaviorally and socially independent at hatching or birth. Human development has moved strongly toward the altricial side, and our brains, in particular, undergo a great deal of their growth following parturition, rather than before—in part, it would seem, as an accommodation to the size of the birth canal. Indeed, even our near relative *Homo erectus* had a much more simian pattern of precocial development compared to our altricial pattern.[25]

JUVENILE LIFE EXPECTANCY SPREADS UPWARD

Many gerontologists have hinted at the possibility of connecting the low death rate characteristic of the human juvenile stage to later stages. Mervyn Susser even documents the spread of low death rates: "as the infant mortality of each successive birth cohort declined, so, equally regularly, the age-specific mortality of each of these cohorts at successive ages declined."[26] Leonid Gavrilov and Natalia Gavrilova suggest that "the prospects for prolonging human life . . . [are linked to the] practicable task of gradually reducing the risk of death at each age" and come very close to speculating on prospects for neoteny: "It is [after all, only] at age 10–15 that we see the beginning of the age-dependent growth in total mortality and mortality from a number of 'endogenous' causes, and the first signs of degenerative change appear (for example, atrophy of the thymus)."[27] Richard Cutler comes even closer to attaching neoteny to prolonged longevity, suggesting that an "increase in maximum life-span potential appears to be possible only by decreasing the overall aging rate, which in turn would result in a correspondingly uniform prolongation of health," and later, while discussing the effects of hypophysectomy, suggesting "that some aspects of aging can not only be slowed down but actually are reversed to a juvenile level."[28]

To whatever degree neoteny is afoot in human evolution, it does not seem to have produced a permanent, morphological juvenile (a human equivalent to the mud puppy) so much as it has *rejuvenated* individuals as they move into later stages of adulthood. Neoteny has resulted in the extension of the juvenile virtues of vim, vigor, and vitality into sexually mature, adult life. And, because the juvenile stage is the healthiest and least fragile stage, if not the strongest stage of a lifetime, neoteny has resulted in the prolongation of life itself.

SLOWING THE DECLINE OF ORGAN SYSTEMS

Assuming that the juvenile stage epitomizes the optimal condition of interacting organ systems, the stretching of these systems from the juvenile stage into later stages would promote longevity. Several systems built by stem cells and containing stem, reserve, and cache cells within their borders come to mind as possible candidates for spreading juvenilization, but the lymphopoietic and hematopoietic systems would seem to have the greatest potential for promoting longevity.

Of course, the complexity of immunity cannot be exaggerated. Investigators have even recommended that an "approach [to] immunology via the science of chaos and fractals . . . would be more appropriate than classical methodology."[29] But of all the body's systems, the immune system would seem to have the broadest reach across the many dimensions of aging. From the endocrine system (the thymus-pituitary and thymus-adrenal axes) to the nervous and neuroendocrine system, from gestation to morbidity, nothing seems to escape the immune system. And, although the immune system itself is integrated and interdependent, several of its more salubrious functions peak in the juvenile stage, suggesting how extending the juvenile condition of the immune system to older stages of a lifetime might extend life itself.

Aging in the immune system is linked to a decline in the ability to produce protective antibodies in response to immunization, immuno-senescence, the age-related breakdown of immuno-surveillance, unresponsiveness to infection, and increased incidences of autoimmune disease and certain cancers.[30] Typically, one assumes that the old immune system is simply exhausted. The body has, after all, been exposed to viruses, bacteria, parasites, food, and one's own, or self-molecules, continuously and unavoidably over a lifetime. But those who have lived to old age have immune systems that have reacted successfully to many of these stresses and produced immuno-reactive and memory T-cells capable of coping with many antigenic stimuli. The problems of immuno-senescence are subtler than "exhaustion."[31]

Major age-related diseases such as atherosclerosis, dementia, osteoporosis/osteoarthritis, and diabetes may erupt as a consequence of a progressive imbalance between the two parts of the immune system: innate and acquired. Innate immunity, associated with inflammation and macrophages in tissues, natural killer (NK) cytotoxic activity, chemotaxis, phagocytosis and complement, is largely unaffected or may even be enhanced by age. On the other hand, acquired or clono-typical immunity, featuring the specificity of B- and T-lymphocytes, deteriorates with age as if high levels of mitotic activity have resulted in mitotic senescence and clonal exhaustion.[32] In particular, humoral immunity (antibody production) is impaired, presumably when fewer pro- and pre-B cells in the bone marrow produce less recombination-activating gene-1

(RAG1) mRNA required for the rearrangement of immunoglobulin (Ig) heavy and light chain gene segments.[33] Moreover, the decline in the antibody-producing response and delayed-type hypersensitivity with age may be due to the loss of T-helper (TH) cells or to their failure to "home" to the spleen.[34]

Thymic Involution

Other features of aging, such as increased susceptibility to certain bacterial (*Escherichia coli, Streptococcus pneumoniae, Mycobacterium tuberculosis, Pseudomonas aeruginosa*) and viral (Herpes virus, influenza virus) infections seem to be consequences of age-related thymic involution and, consequently, the failure to produce virgin T-cells. Changes in the immune system thought to be thymic-dependent include T-cell unresponsiveness, failure to proliferate, declining proportions of lymphoid cells to total bone marrow cells,[35] and failures in conducting transduction pathways affecting T-cell activation by monocytes and macrophages.[36] Indeed, the strongest evidence supporting the importance of these changes in immuno-senescence is the absence of these changes or their moderation in centenarians—those who survive to extreme old age have adequate immune responses.[37]

In mice, prior to puberty lymph nodes and spleen, as well as the thymus, decline in relative weight, while after puberty, only the thymus continues to involute and influence changes elsewhere. In human beings, thymic involution begins in the infant, peaks in the first twenty to thirty years, and continues until middle age. Thymic involution (specifically *Altersinvolution,* as opposed to accidental, gestational, and lactational thymic involution) results in the reduction of thymocyte production by 3 percent to 5 percent per year and hence a decline in concentration of thymic-dependent T-cells in peripheral lymphoid organs. Following middle age, the rate of reduction slows down to less than 1 percent per year.[38] Specifically, thymic involution induces a decline in T-cell receptor excision circles (TREC) concentrations in circulating T-cells (both CD4+ and CD8+ varieties).[39]

Speculation on the mechanism of thymic involution runs in three directions, but each involves loses of reserve, stem, or transit cells: (1) A failure of renewal in thymic epithelial cells creates changes in the thymic microenvironment, specifically a diminution of the cytokine interleukin 7 and hence a failure of thymopoiesis and T-lymphocyte dysregulation;[40] (2) stromal (thymic epithelial) changes have no effects on the impairment of thymocytes (parenchyma),[41] but thymopoiesis is disrupted by mutant genes early in T-cell development[42] or by an increase in the frequency of programmed cell death (judged cytologically through instances of apoptosis) coupled to a diminution in cell division resulting in a reduction of thymic cortical cellularity and hence in the release of impaired T-lymphocytes;[43] (3) both stromal and parenchymal effects[44] running through multiple steps:[45] the microenvironment, with diminished numbers of aged thymic stromal cells is less efficient in supporting the repopulation of peripheral thymoyctes compared with that of young organisms, and a block in the development of T-cells diminishes the rate of proliferation in peripheral T-cells in older organisms.[46] Presumably, under the impact of neoteny, enlarging the population of stem and cognate cells and, hence, transit cell populations, would moderate thymic involution and ameliorate deleterious effects.

Dysregulation of Hematopoiesis

Multiple lesions also seem to be involved in the dysregulation of hematopoiesis in the elderly. In particular, reduced production of granulocyte/macrophage colony stimulating factor (GF-CSF) may result in reduced numbers of progenitor cells in bone marrow. Aging is also associated with reduced numbers of committed hematopoietic progenitor cells, cells with a high potential for auto-transplantation, and cells available for mobilization and repopulation following treatment with cytokines.[47]

On the one hand, the bone marrow of elderly people is hypo-cellular for hematopoietic stem cells, particularly CD68 positive cells associated with the macrophage population. Moreover, aged mice produce less vascular endothelial growth factor (VEGF) than younger mice and, hence, experience delayed wound re-epithelialization, collagen accumulation, and angiogenesis.[48] On the other hand, functioning antigen presenting cells (APCs) may be well maintained in the elderly.[49] Indeed, proliferative activity of cells in the primitive hematopoietic compartments (of mice) is greatly reduced with age, but increases in relative and absolute numbers of autonomously cycling SR cells may compensate for the loss.[50] In any event, the maintenance of the juvenile condition of hematopoiesis into adult life is more likely to be salubrious than detrimental.

CELL-LEVEL CANDIDATES FOR JUVENILIZATION

Abundant data from the annals of longevity genetics suggest that aging is associated with increased rates of stress-induced programmed cell death (PCD) and the death of senescent somatic cells. Stress—in particular, oxidative damage induced by reactive oxygen species (ROS)—is associated with cell death, neurodegenerative disease, retinal degeneration, cardiovascular disease, and vulnerability or fragility. Precisely these stress-induced processes would have to be the targets of neotenous reform.

On the other hand, caloric restriction (CR) increases expression of SIRT1 in rodent tissues including brain, visceral fat pads, kidney and liver, protecting cells from stress-induced apoptosis, while serum from CR animals promotes proliferation of cells in vitro and the attenuation of stress-induced apoptosis. Thus, the induction of SIRT1 expression by CR could promote the survival of irreplaceable cells and thereby extend life span. Similarly, the oxidative stress response is offset by targeted mutation of the mouse $p66^{shc}$ gene, a cytoplasmic signal transducer, thereby inducing stress resistance and prolonging life span.[51]

Optimally, SR stem-cell dynamics sustain steady-state tissues throughout a lifetime, renewing their own members while replacing precursor and transit cells lost in normal physiological turnover. Juvenilization may result from increased numbers of SR cells, but recruitment of stem cells from non–stem-cell sources would seem unlikely for several reasons. With the possible exception of hematopoietic bone marrow cells,[52] adult SR cells have little if any ability to transdifferentiate or change their determination.[53] Adult SR cells may represent the base level of differentiation, but they are not free to deviate from their already-determined course of differentiation under normal circumstance. Juvenilization or neoteny via the continued presence of SR cells and cognate cells within tissues at something approaching their juvenile numbers is more likely, therefore, to result from increased numbers of original SR cells rather than transdifferentiation or recruitment from non–stem-cell sources.

When trying to imagine where additional SR cells come from, researchers interested in the genetics of aging or the telomere story gravitate toward the possibility of increased cell division. Were telomerase expression prolonged in cells and telomere length stretched as a result, cells may very well divide for longer periods of time, live longer, and increase prospects for replenishing worn out or damaged body cells. In particular, enhanced telomerase activity might well preserve vital cells that would otherwise die prematurely—for example, neurons in the substantia nigra whose death is correlated with Parkinson's disease, and β-cells in pancreatic islets whose disappearance is associated with diabetes.

The argument on behalf of extending cells' ability to divide is, however, problematic. In the first place, some notorious cancer cells are known to divide indefinitely. They are said to be immortal, although they kill the organism bearing them and die with their host. But even normal cells dividing to excess would pose problems for healthful regulation: how is an adult organism's non-expanding body to accommodate excess cells? Would the organism burst from all the cells produced by excess division? Alternatively, the organism might evolve a mechanism for getting rid of the excess—thereby undoing the advantage gained by enhanced cell division!

The possibility of mutations accumulating in cells dividing to excess is yet another problem for enhanced proliferation. Ordinarily, it would seem, telomeres and the activity of the enzyme telomerase tailor the number of cell divisions performed by normal cells and bring about mitotic senescence before the cell has accumulated a deadly dose of replication errors. Excessive divi-

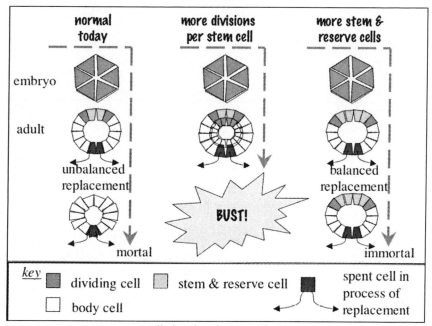

FIGURE 6.1. More stem cells lead to longer life.

sion might upset this balance, promoting mutation along with the delay in mitotic senescence in somatic cells to the detriment of normal cell function.

For these reasons, instead of more cell divisions, *the acquisition of more stem and cognate cells* would seem a more likely scenario for neoteny and prolonged longevity. The change that might promote our longevity is increasing the number of SR cells and, hence, maintaining adult tissues longer in a juvenile state.

The neotenized adult, thus, would have a normal number of TA and differentiated cells appropriate for the juvenile, but a greater proportion of tissue cells would belong to the SR cell compartment. In effect, the neotenized adult would have the potential to maintain and restore itself at the juvenile level well into adult life as a result of the excess of SR cells.

Diverting Primordial Germ Cells to Somatic Compartments

The lynchpin connecting neoteny with an expanded life span would seem to be the embryonic addition of SR cells to tissues, thus tipping the scale in favor of extended cellular replacement in adult organisms. Given the constraints discussed above on increasing the rate of cell division, the additional SR cells cannot simply be produced in situ by promoting cell division. Rather, pluripotential stem cells—cells capable of differentiating into a large array of cell types—from a preexisting embryonic compartment would have to be recruited into SR compartments of various adult somatic tissues.

One embryonic compartment comes to mind as the most likely source of these somatic SR cells. In fact, it is the only compartment that could suffer a loss of cells without the organism paying too high an anatomical or physiological price—and that compartment is the normal source of primordial germ cells (PGCs)!

After tissue culture, the progeny of PGCs are pluripotential on a scale matching that of ES cells. Indeed, GS cells seem to have no limits for differentiation after introduction into blastocysts or in vitro, including successful maturation into *functional* germ cells.[54] In vitro, "[t]here is a lack of markers that can suitably distinguish between ES cells and PGCs."[55]

The idea of stem-cell reallocation changing the roles of GS cells is not as alien as it might sound on first hearing. Indeed, it is frequently rehearsed in the scientific and popular literature. In principle, ES cells are notorious for displaying pluripotentiality.[56] Evolution might very well reallocate a pluripotent variety of stem cells in embryos from one generalized compartment, consequently seeding different adult tissues with additional SR cells.[57]

NEOTENY AND THE GERM LINE

Neoteny could affect the germ line directly by altering the time or rate of its appearance. Indeed, one is not surprised that in the polyembryogeny of the wasp, *Copidosoma floridanum,* members of the soldier caste of developmentally arrested larvae are not only sterile, but their primordial germ cells are reallocated to their fertile sisters.[58] In human beings, neoteny, in contrast to progenesis, would be expected to delay the appearance of the germ line or retard its rate of emergence, possibly resulting in fewer germ cells and a drop in fecundity. The question is, thus, when do germ cells first appear in human embryos and can their appearance delay or retard the course of development?

Ever since August Weismann divorced germ and somatic lineages (see chapter 3), developmental biologists have assumed that the germ line separates from all somatic lines during cleavage. In mollusks and annelids exhibiting spiral cleavage, the germ line is isolated during cleavage,[59] and in *C. elegans,* the separation of germ and somatic lines of cells occurs at the fourth cleavage.[60] Likewise, in insects, early cleavage nuclei reaching the posterior polar cytoplasm become determined for the germ line. Germ cells could hardly appear any earlier, but in cnidarians, tunicates, and flatworms, much like plants, germ and somatic lines are never completely separated except in adults prior to germ-cell differentiation, and in the preponderance of metazoans and virtually all plants, algae, protozoans and fungi, the soma gives rise to the germ line![61]

Until recently, vertebrates were thought to fall in between these extremes. In amphibians, birds, and mammals PGCs, identified by large size, dark staining, spherical or ellipsoidal shape, high nucleo-cytoplasmic ratios, and perinuclear concentrations of mitochondria, were not found until well after cleavage,[62] even if determination happened earlier under epigenic influences.[63] In anuran blastulas, inducers of PGCs would seem to be present in the region of vegetative endoderm, since the removal of this region eliminated PGCs and its reposition moved the site of germ-line determination.[64] Likewise, in urodeles, PGCs normally appeared in the marginal zone of the gastrula's presumptive lateral plate mesoderm, but presumptive PGCs could be induced in the animal pole by contact with the vegetal hemisphere.[65] Similarly, in chicks, removal of the germinal crescent at the junction of the blastoderm and anterior yolk eliminated PGCs and resulted in sterility.[66]

Initially, germ-line determination in mammals appeared to be even later than in other vertebrates, since killing blastomeres in early mammalian blastocysts or adding additional blastomeres to blastocysts failed to affect germline determination.[67] This result is not entirely surprising, however, since malignant mouse teratocarcinoma cells injected into mouse blastocysts in small numbers contribute to various cell lines, including the germ-cell line.[68] Thus, mammalian blastomeres seem undetermined early in development (they are regulative), and embryonic cells seem generally capable of forming all tissues, including the germ line.[69] Indeed, human GS cells in vitro are obtained from gonadal ridges and mesenteries of aborted fetuses as late as 5- to 9-weeks post-fertilization.[70]

Views on determination of germ cells in the early mammalian embryo have shifted dramatically, however, since the isolation of GS cells in vitro. Now, it is held that "[i]n the mouse, germ cell competence is induced at embryonic day 6.5 in proximal epiblast cells by signals emanating from extraembryonic ectoderm."[71] Primordial germ cells in the proximal epiblast of the blastocyst then migrate to the posterior endoderm where they can be found 8 days post-coitus.[72] Similarly, in the human embryo, PGCs first appear among yolk-sac endodermal cells in the vicinity of the allantoenteric diverticulum at the notochordal process stage (Stage 7, about 0.4 mm in diameter) 15–17 days post-ovulation. Two days later, the embryo has advanced to the neural groove stage (Stage 8, 1.0–1.5 mm length) with mesonephric ridges, a dorsal mesentery, and the intra-embryonic mesoderm otherwise split by the intra-embryonic body cavity; and at 28 to 32 days post-ovulation (Stage 13 embryos, 4–6 mm in length) the PGCs begin their perilous journey in the dorsal mesentery of the hindgut. Then, during the fifth week post-ovulation (Stage 15, 7–9 mm long), PGCs begin colonizing the indifferent gonadal ridges, and by the sixth week (Stage 17, 11–14 mm in length), PGCs have given rise to TA cells—gonocytes or precursors of germ cells—and, following myriad divisions, their descendents settle down in the fetal gonad as primary oocytes or dormant spermatogonia (see chapter 5).

The germ line's induction in the epiblast, thus, could make it a target for neoteny either through delaying or retarding development. It is not too hard to imagine how primitive, pluripotential cells in the proximal epiblast could be pushed away from PGC determination toward any of a variety of other destinations. After all, hematopoietic stem cells of the embryonic yolk sac are derived from the same cell pool as PGCs. Under evolutionary pressure, PGC determination might be delayed, while competitive somatic determination swallows up an increasingly large share of the primitive, pluripotential epiblast cells. Such a reallocation of potential PGCs to somatic-cell lineages might also cause a diminution in the number of gonocytes, resulting in diminished fecundity.

Neoteny, thus, is not ruled out by the timing of germ-line determination. One can even imagine a great, if reversed, neotenous horse race taking place in human evolution: which adult cell line develops slowest, germ or somatic? Greater recruitment of somatic stem cells might result in a diminution in the number of germ cells in the embryo and hence reduced fecundity in the adult.

FECUNDITY IS DECREASING

The archives on fecundity could provide several sorts of evidence on germ-line reallocation. Perhaps the first things one would expect to find in a species undergoing neoteny would be a decrease, as well as a delay, in reproduction.

This is not to say that fitness would also decline. On the contrary, greater parental investment accompanying a switch from precocial to altricial development might very well compensate for diminished fecundity.

And, indeed, at the same time that life expectancy has increased, fecundity has decreased. According to the U.S. Census Bureau, the pace of population growth peaked in 1963–1964 and has since declined.[73] From a worldwide fertility rate of 5.0 thirty years ago, "[t]otal fertility rates currently equal 2.1, 1.3, and 1.5 in the U.S., Japan, and the EU [European Union], respectively."[74] What is more, "[t]he slowdown in the growth of the world's population can be traced primarily to declines in fertility. In 2002, the world's women, on average, were giving birth to 2.6 children over their lifetime. This was less than one-half of a child more than the level needed to assure the replacement of the population. . . . Census Bureau projections suggest that the level of fertility for the world as a whole will drop below replacement level before 2050."[75]

Not surprisingly, in the 2004 annual report of the U.S. Social Security Trustees, the ultimate values for the total fertility rate (children per woman) that are assumed to be reached within five to twenty-five years, are 1.95 (most-

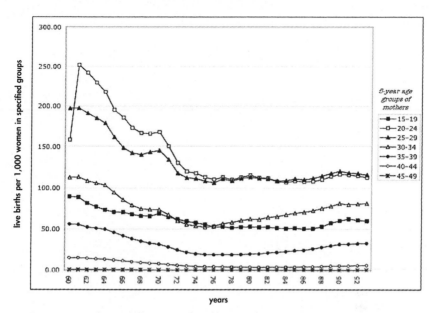

FIGURE 6.2. Birth rates by age of mother in the United States, 1960–1993. (From National Center for Health Statistics, *Vital Statistics of the United States*, 1993, vol. 1, [Hyattsville, Md., 1999].)

likely case scenario), 2.2 (worse case scenario), and 1.7 (best case scenario),[76] in other words, at or below replacement.

A general drop in fecundity was neither restricted to women in the United States (12 percent from 1990 to 1999), nor was it of recent origin.[77] The drop is virtually species-wide![78] Indeed, "[b]y the end of the nineteenth century it was common knowledge that fertility levels were falling in many Western countries and there was a presumption that birth rates would stabilize at lower levels."[79] Moreover, "fertility decline in the Third World is not dependent on the spread of industrialization or even on the rate of economic development. It will of course be affected by such development . . . [b]ut fertility decline is more likely to precede industrialization and help bring it about than to follow it."[80]

Two interesting, if subtle, points relevant to neoteny emerge from the U.S. Census Bureau data: (1) the pattern of fecundity has shifted toward older women, reflecting the trend toward delayed childbearing for the population of women in the United States; and (2) more "women with fewer children as well as those bearing children late in life live longer post-reproductive lives. . . . [Incidentally,] husbands have effects that are similar to those of their wives during the latter marriage cohort."[81]

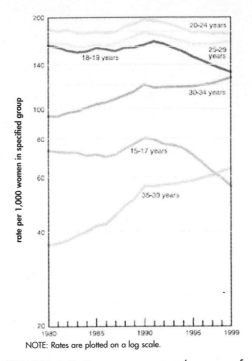

NOTE: Rates are plotted on a log scale.

FIGURE 6.3. Pregnancy rates by age of mother in the United States, 1980–1999. National Vital Statistics Reports, vol. 52, no. 7, 2004, p. 2.

Thus, while younger women (20–24 and 25–29) are showing the greatest decline in childbearing, older women (30–34 and 35–39) are showing an increase in the number of pregnancies and children born. Such a delay is precisely the sort of thing one would expect of a species undergoing neoteny: an apparent advantage—prolonged life and hence an opportunity for increased biparental investment—coupled to delayed reproduction and hence prolonged robustness. Were we fruit flies, no one would have difficulty accepting this interpretation. Indeed, when Michael Rose selected breeding flies from those producing eggs later and later in their lifetime he succeeded in isolating long-lived flies.[82] A linkage, thus, would seem to be established between late fecundity and long life. What remains is connecting delayed fecundity and germ cell reallocation to SR cells associated with long life—no small order but well within the competence of empiricism.

In the case of Rose's experiments with *Drosophila,* selection was totally artificial, that is, a technician selected the last flies contributing eggs to the

donors born in the	1950s	1970s	P
median sperm concentration (X one million / ml)	98	78	
10th to 90th centile	38.6–218.4	21.0–166.4	0.002
total number sperm (X one million)	169.7	129	
10th to 90th centile	52.3–503.9	29–325.8	0.0065

FIGURE 6.4. Semen quality of birth cohorts.

population. In the case of human beings, the selection may be more natural. In particular, a premium may be placed on grandmothers (the so-called "grand-mother effect") capable of promoting the reproductive success of their children and the survival of their grandchildren. Indeed, an analysis of premodern populations of both Finns and Canadians using complete multigenerational demographic records shows that "women with a prolonged post-reproductive life span have more grandchildren, and hence greater fitness."[83] The anthropologist Kristen Hawkes might be discussing neoteny as much as the grandmother effect when she suggests that "females ageing more slowly in physiological systems other than their ovaries could help [their offspring and heirs even] more."[84]

Females are not, however, the only members of the species with changing patterns of reproduction, although spermatogenesis and sperm quality rather than fecundity is the relevant variable for males.[85] Although "[r]eports on decreased sperm motility, semen volume and changes in sperm count are contradictory,"[86] a profound decrease in "semen quality" seems to have taken place in Europe when fresh ejaculates collected in the 1970s are compared to those from the 1950s. Both the median sperm concentration and the total number of sperm per ejaculate were reduced significantly in samples obtained twenty years apart despite the donors in the 1950s having been older than donors in the 1970s (means 34 and 20 years, respectively; see figure 6.4). These reduced sperm counts suggest that either fewer primordial germ cells entered the pathway toward spermatogenesis or fewer completed the process and differentiated as mature spermatozoa.[87] Germ-line productivity in both women and men, therefore, may be inversely correlated with longevity.[88]

Of course, many social and environmental factors have been offered as explanations for the parallel trends in fecundity and death rates, but evolution might also be at work, shaping fecundity and life span at a fundamentally biological level and profoundly changing our species. While for a long time life-history theoreticians have argued that "[m]any life-history models have increased adult mortality as a cost of reproduction,"[89] the shoe may now be on the other foot, so to speak, and the costs may lie in the opposite direction. Indeed, frailty and longevity may rest on antagonistic functions of germ and somatic lines.[90] *Homo sapiens*, rather than *Drosophila,* may well be the species epitomizing the prolongevous advantage of delayed fecundity, and the slowing trend in our rate of maturing and aging may well epitomize neoteny among vertebrates.

How, then, might neoteny have evolved in recent times? Neoteny's evolution may well have sprouted from niche construction—that organisms inevitably cause changes in their environment that, in turn, affect their own fitness.[91] This notion is credited to Richard Lewontin, but it seems to reflect James Lovelock's Gaia[92] and V. I. Vernadsky's biosphere.[93] Whether organisms modify their environments and environments subsequently select organisms, or organismal-driven changes in the environment persist and select members of subsequent generations, a species' evolution is driven at least in part by the impact of the species on its niche.

Today, it is fashionable to mention global warming, desertification, and the destruction of tropical rain forests as examples of our impact on our environment and hence on ourselves, but a large part of our daily activities would seem to reflect back on that part of our environment with implications for neoteny. For example, one can hardly doubt that through moderate exercise, healthy foods, safe sex, and so on, we not only promote the extension of youthfulness, but precisely these practices select for those who benefit from them. If moderate stress preserves or even expands our reservoirs of SR cells, our progeny in the next generation can be expected to have larger reservoirs and more readily expandable populations of these cells than we have. Virtually everything concerning mortality and life expectancy demonstrates a parallel between our biology and our potential to select youthful traits. Our prejudices cannot help but come back to haunt us: The ghost of our cultural investment in youthfulness is neoteny.

IN SUM

In retrospect, the bagpipe model for lengthening life span discussed in chapter 2 is entirely consistent with the preemption of presumptive PGCs for service as SR cells and spreading juvenile "virtues," especially those of the immune system, outward beyond the juvenile stage to later adult stages. The premise of the neoteny hypothesis is that we are living longer because our adult stem cell populations have been enlarged over the course of our recent evolution. Thus, we live longer now than we have in the past because these cells, and hence TA cells, are available to us for increasingly long periods of time.

The question "where did the additional stem and cognates cells come from?" is tentatively answered by the recruitment of PGCs. This possibility is consistent with evidence for a contemporary drop in fecundity and with evidence for the induction of germ and stem cells from the same embryonic compartment. If neoteny is slowing development generally, potential stem cells might even remain pluripotent for a prolonged period, and, hence, be vulnerable to recruitment into somatic SR cell compartments and away from the germ-line compartment of PGCs.

Afterword

The constant effort to dispel this darkness, even if it fail of success, invigorates and improves the thinking faculty.
—Thomas Malthus, *Essay on the Principle of Population*

We love machine-gun massacres in movies, but death from old age seems somehow unnatural and horrifying.
—David Lovibond, "No Way to Grieve"

If we are to conceive Man as separate from nature, then Man does not exist. This recognition is precisely the death of Man.
—Michael Hardt and Antonio Negri, *Empire*

In the preface, you may remember, I suggested that science is like a whodunit, a detective story where a crime is uncovered and a mystery is solved. As things turned out, the evolution of death is not the crime I thought it was. Rather than a corrosive force shortening life, death's evolution turned out to be a salubrious force lengthening life!

It seems that our life span has lengthened, because our own intrusions in our environment (such as improved sanitation, nutrition, and medicine) have made our niche more wholesome than it ever was before, and, as it turns out, the very things that shape our longevity are the very things that select individuals for delayed death. Instead of dying while still near our peak, we are living longer than ever before. Increasingly, selection is choosing individuals with a greater potential for life. Consequently, our species is in the process of optimizing death! In other words, *death is becoming more efficient and less costly.*

Death, it would seem, does not turn us into corpses as quickly as it once did, but the mechanism for death's delay remains a subject of speculation. My guess is that death's evolution is dominated by neoteny—the emergence of sexual maturity in juvenile morphs coupled to a tendency toward slower anatomical development—and we are living longer because our juvenile robustness is preserved later into adulthood. If, as I suggest, parsing germ cells into compartments of somatic stem cells and their cognates has expanded our stem-cell reserves, our sources of cellular renewal would be maintained at

151

youthful levels well into adult life, and our youthful vigor would be preserved well into our mature years. We might, thus, delay reaching the time our cellular resources are disabled, inadequate or exhausted and, hence, put off meeting our death.

But here the plot thickens, and *The Evolution of Death* poses two new mysteries. First, why has death's evolution escaped notice by gerontologists? Second, where is death's evolution presently heading? What will happen to us if death's evolution continues at its present pace or even accelerates? I won't pretend to solve these mysteries, but let me begin this afterword with a critique of gerontology and proceed to project some possibilities for humanity in the future.

HOW DEATH'S EVOLUTION ESCAPED
THE GERONTOLOGIST'S NOTICE

THE GAINS

Gerontologists seem widely in agreement that human beings are living longer, which is to say that our median age at death is greater now than ever before. But many gerontologists still cling to the notion that we die after a definitive human life span and life's ultimate boundary has not changed. The alternative notion that our life span can change and our death can evolve seems unacceptable if not loathsome to these gerontologists. Their problem is a commitment to aging genes.

Frankly, I can sympathize with many gerontologists mired in biology's prevailing gene paradigm. Nothing has prepared them to think about the prolongation of longevity, no less the evolution of death in epigenetic terms. I am, however, much less sympathetic with close-minded gerontologists who hound and abuse those struggling to understand longevity in new terms, notably the bio-gerontologist Aubrey de Grey. Alternatives to genetics, such as the epigenetic roles of mitochondria, DNA methylation, the reshaping of chromatin, and changes wrought by transposons and other "junk" DNA must be examined before we can claim to understand the inheritance of aging and the evolution of death.

The notion that human beings are allotted three- or fourscore and ten years has ancient roots that remain stout today even if they are now proving to be of clay. Notwithstanding biblical claims for Methuselah and his kin's extraordinary longevity, The Booke of Psalmes tells us that old age arrives at threescore and ten years and we're lucky to survive fourscore years and ten.[1] This conception of our lifetime was not only endorsed during the Renaissance[2] but reinforced throughout the Enlightenment. As the political economist

Thomas Malthus told us at the end of the eighteenth century, "it may be fairly doubted whether there is really the smallest perceptible advance in the natural duration of human life since first we have had any authentic history of man."[3] Malthus supported his conviction with the prevailing prejudice against evolution: "With regard to the duration of human life, there does not appear to have existed from the earliest ages of the world to the present moment the smallest permanent symptom or indication of increasing prolongation."[4]

Still, Malthus, who died at 68+ years,[5] well beyond the median for his cohort, had doubts. He did not cite mortality statistics for the British Isles in *An Essay*, but in *Summary* (first published in 1830 in the *Bibliothèque Britannique,* vol. 4) Malthus used M. Muret's data for Geneva and his "calculations, . . . [according to which, in] the eighteenth century, the probability of life [that is, the age to which half of the born lived] had increased to 27.183, twenty-seven years and two months; and the mean life [the average number of years due to each person] to thirty-two years and two months." Other statistics attributed to M. Muret showed the mean life in Lyonois was "little above twenty-five years; while in the Pays de Vaud, the lowest mean life, and that only in a single marshy and unhealthy parish, is 29.5 years, and in many places it is above forty-five years . . . [While in] the parish of Leyzin . . . the probability of life was as high as sixty-one years."[6] Malthus was also (presumably) aware of alleged "exceptions to the rule"[7] and rare individuals who actually lived beyond 100 years.[8] Unlike so many of our contemporary commentators on life span, Malthus granted the possibility that "by an attention to breed, a certain degree of improvement, similar to that among animals, might take place among men . . . [inasmuch as] size, strength, beauty, complexion, and perhaps even longevity are in a degree transmissible."[9]

How then would Malthus have dealt with our reaching a median age of death about 80 and the well-documented record of Jeanne Calment who died in 1997 at the age of 122+ years?[10] Would he have pointed to good breeding and, however reluctantly, acknowledged the possibility of evolution playing a part in increasing longevity? He would certainly have added the caveat, "It has appeared, I think, that there are many instances in which a decided progress has been observed, where yet it would be a gross absurdity to suppose that progress indefinite."[11]

And, no doubt, Malthus would have reacted skeptically to predictions based on British and Swedish mortality data that "the mean life-span [of a 30-year-old male is] . . . about 350 years"[12] and the possibility that a "rollback to the robust physiology of your late teens or early twenties would . . . push your Expected Age at Death up to around 700–900 calendar years . . . [Moreover,] if we can eliminate 99% of all medically preventable conditions that lead to natural death . . . your healthy life span, or health span, should increase to about 1100 years."[13] Surely, Malthus would have rejected the claim made by

Aubrey de Grey that teenagers in 2004 already have a life expectancy of 1000 years[14] and "that the AVERAGE life span of those in wealthy nations born in 2100 will be 5000."[15]

Nevertheless, today it hardly seems necessary or prudent to deny the possibility of life span extension. And, of course, long life does not come alone: In both birds and mammals, including us, long life is associated with delayed childbearing and low fecundity.[16] If these trends in longevity, childbearing, and fecundity were to continue, it would seem, we would rapidly become a species of negligibly aging, slowly reproducing organisms with an indeterminate life span!

THE STRAINS

The current spurt in life's prolongation among individuals living in the developed and developing worlds has already strained the institutions charged with hygiene, sanitation, and public transportation, but the greatest strains are in institutions supporting social welfare and health care of aged citizens. The proportion of aging citizens in advanced countries is bulging and is about to bulge even more. Beginning in 2011,when baby boomers (people born post-World War II) begin to reach full pension age in the United States, means supporting social pension funds, designed for entirely different demographics, will be strained, and, before the end of the century, these means will be unable to cope with the bumper crop of retirees at contemporary levels of support.

In the beginning, in 1889, when Germany Chancellor Otto von Bismarck decreed that a pension would kick in for all workers in trade, industry, and agriculture at the age of 70 years, he was motivated by the demands of labor and his determination to defeat an incipient socialist revolution. However, few workers lived to 70 at the time. In fact, as late as 1911 (when complete records became available for women and men) in Germany, the median life expectancy was 51–52 years for women, 46–47 years for men.

By 1901, Austria, Italy, Sweden, and the Netherlands had followed the German example and created their own forms of state-guaranteed pensions, but the age of retirement was still far above the median life expectancy (Austria [1907]: 51 years for women and 44–45 years for men; Italy [1901]: 49–50 years for women and 34–35 years for men; Sweden [1901]: 51 years for women and 47–48 years for men; the Netherlands [1901]: 51–52 years for women and 48–49 years for men[17]). The United States got into the act during the height of the Great Depression, passing its Social Security Act in 1935 and establishing its own old-age insurance system financed by a payroll tax on employers and employees. At that time the median life expectancy in the United States was 54 for women and 46 for men!

What has happened since then? By the 1950s,[18] the median life expectancy surpassed the age of retirement in Austria, Germany, Italy, Sweden, the Netherlands, the United States, and Japan. Thus, the expectation that more people would always pay into the pension fund than would withdraw from it became shaky. By the 1980s the costs of social pensions had become staggering: 32 percent of the gross national product of Sweden; 25 to 30 percent in Belgium, Denmark, France, and the Netherlands; 20 to 25 percent in Austria, Germany, Ireland, Luxembourg, and Norway; 18 percent in the United Kingdom; 13 percent in the United States; and 11 percent in Japan. Indeed, by the advent of the twenty-first century, the median life expectancy for populations in most developed nations was approaching 80 years and the number of pensioners over 65 had soared.

This upward trend in costs is likely to continue in industrialized nations. Above all, the proportion, if not the number, of aged in the population is likely to increase in the twenty-first century, and women, who will make up an increasing proportion of the elderly—simply because they live longer—will demand pensions equal to those of men, something women have rarely received. The cost of health care for the elderly will also increase disproportionately, especially given improvements in costly medical technology.

Of course, several cost-cutting adjustments have been tried through legislation over the years, some of which, or some combination of which, may yet staunch the flow of cash. The age of retirement has skidded back and forth, and retirement has changed from compulsory to non-compulsory under some circumstances. In Europe, under pressure to create jobs for young people, the retirement age has moved downward to between 55 and 66 for women and 60 and 67 for men, while in the United States, under pressure for stability in long-term financing, the age of retirement is moving up from 65 to 67.

And several more adjustments can be anticipated before the institution of governmental-guaranteed pensions is bankrupt. For example, an individual beyond the retirement age might be allowed to receive a partial pension while working at reduced hours, thereby making a gradual transition to full retirement while opening up a slot in the workplace and increasing the tax base. Limits might also be lifted on what pensioners can earn without reducing their pensions, as long as the excess earnings are taxed as regular income. And, of course, several schemes have been floated to "privatize" the state pension. Private-sector pensions, are, after all, financed through capitalization, but privatizing state pensions would destabilize the market by creating an unbearable demand for high cash yields on investments. Moreover, replacing the protection inherent in cost of living raises currently attached to state pensions with the promise of increased gains, interest, and dividends is not likely to produce peace of mind. In effect, demographics may have defeated state pension schemes and the aging population can no longer be guaranteed the fruit of its labor. But, then again, maybe not.

In the United States, at least, the doom and gloom predictions for Social Security solvency seem to grow from political machination and maneuvering rather than actuarial statistics. Indeed, the entire problem may be wiped out by further changes in fecundity.[19] In this vein, the Social Security Trustees, which considered "all the demographic, economic, and program factors that affect income and expenditures . . . [recently concluded that] projected annual balances for the Social Security program (income minus costs) are somewhat improved for years after about 2045."[20] However, "[s]eparately, the OASI [Old-Age and Survivors Insurance] and DI [Disability Insurance] funds are projected to have sufficient funds to pay full benefits on time until 2044 and 2029, respectively. By 2078, however, annual tax income is projected to be only about two-thirds as large as the annual cost of the OASDI program."[21]

Some adjustments, therefore, will be necessary, but a modest fix to keep Social Security solvent through taxation would not seem too drastic and, certainly less costly than privatization.

> For the trust funds to remain solvent throughout the 75-year projection period, the combined payroll tax rate could be increased during the period in a manner equivalent to an immediate and permanent increase of 1.89 percentage points, benefits could be reduced during the period in a manner equivalent to an immediate and permanent reduction of 12.6 percent, general revenue transfers equivalent to $3.7 trillion (in present value) could be made during the period, or some combination of approaches could be adopted. Significantly larger changes would be required to maintain solvency beyond 75 years.[22]

In other words, we will have to pay for the increased longevity that we experience today, but the payment may not be too great compared to the benefits.

WHERE WILL DEATH'S EVOLUTION TAKE US?

If the most extreme of prospective scenarios for lifetime extension and fecundity remission were to become reality, we would become a species of very nearly sterile individuals having potentially indefinite life spans. How would this kind of life affect living? Would life be endlessly boring for our nonreproductive and endlessly existing descendents or would life be an endless opportunity for creativity and fertilizer for the flowering of civilization? Would individuals be less attached to life or less capable of happiness because life would not be threatened by termination? Would our attachment to life, to projects, relationships, commitments, interests, and memories diminish if we are no longer anxious about death? Would technology cease to advance and our understanding of the universe stagnate or grow if we have time to immerse ourselves in change?[23]

In an age when individuals can expect to live long enough to see the consequences of their actions, will the notion of passing the buck, such as the national debt, to grandchildren cease shaping the nation's priorities, such as spending for warfare? Are people "more likely to be moral when they understand that they will have to face the consequences of their actions in the future . . . [and will longer lives] reduce the tension between the individual and society"?[24]

On the One Hand

Francis Fukuyama, one of President Bush's advisers on bioethics, has recently told us, "[n]o one can make a brief in favor of pain and suffering, but the fact of the matter is that what we consider to be the highest and most admirable human qualities, both in ourselves and in others, are often related to the way that we react to, confront, overcome, and frequently succumb to pain, suffering, and death. In the absence of these human evils there would be no sympathy, compassion, courage, heroism, solidarity, or strength of character."[25]

Earlier, Fukuyama told us, "[f]or by risking his life, man proves that he can act contrary to his most powerful and basic instinct, the instinct for self-preservation."[26] An orthodox Darwinist would have said, "the instinct for reproduction," but the point is the same. Fukuyama goes on to warn the President and the reading citizen that "[f]ear of man's 'lord and master, Death' was a force like no other, capable of drawing men outside of themselves and reminding them that they were not isolated atoms, but members of communities built around shared ideals."[27] In effect, Fukuyama suggests, so much of our sense of moral worth and so many of our values are derived from our commitment to offspring and our awareness of death—if not our fear of it—that a humanity without offspring and death may very well require a different sense of moral worth and utterly new values.

Death, we are led to believe, figures into the evolution of morality like a hand figures into the design of a glove. But does morality make sense without death? On the one hand, death is supposed to be the worse thing that can befall a young, vigorous person. In the United States, for example, death is the ultimate punishment imposed on convicted felons (administered, ironically, by lethal injection in order to avoid the state committing "cruel and unusual punishment"!), while in Europe, capital punishment is banned on the grounds that it is inevitably too cruel to be administered by a civilized society (would that Europeans have the same attitude toward war, "ethnic cleansing," and genocide). On the other hand, death is the object of heroic acts, of great deeds, and potentially an eminent contribution to civil society. Figures achieving immortality in any of these ways are frequently recognized by having their likeness realized in sculptured marble or brass on a path in a public park or in the corner of a museum gallery. But the highest token of esteem and recognition bestowed upon heroes, or at least great leaders (such as kings of European countries and presidents of the United States) is to have formidable instruments of death named after them (such as aircraft carriers).

The fear of violent death is an organizing principle of human relationships—a multipronged mace that cuts in every direction.[28] There are heroes and victims, leaders and followers, masters and slaves, all brought to task by the same threat. Powers to prevent death are conceded to the priest, medicine man, shaman, healer, physician, and surgeon, while powers to impose death on others are bestowed on generals and politicians, constables, magistrates, and judges and juries. Industrialists and businessmen create the jobs that allow the masses to make a living, while workers and farmers create the products and produce that makes life possible, but behind all the enterprise is fear of death.

Death's power to organize values is never stronger than during wartime. Few individuals will acknowledge anything as important as bringing a war to a successful conclusion. War is, of course, the preoccupation of embattled armies, but "total war" dominates civilian life as well: for those struggling to survive on minimum wages, subsistence farming, or handouts from relief agencies, starving masses, oppressed and exploited people, those living under the threat of epidemic disease (including HIV infection), herded into concentration camps or so-called refugee camps (that offer little refuge), and those hunted by occupying powers. For all these individuals, war is all consuming, while those living in relative peacetime, or at least outside a war zone, are largely unmotivated when it comes to finding a substitute for war.

ON THE OTHER HAND

But "[w]hat if it turned out that to be mortal was not an essential condition of our species? . . . What if . . . our species [were] to take evolution into its own hands"?[29]

I have no doubt that humanity can do better than organize itself around death and the fear of death. Endless life holds endless possibilities to innovate, discover, and practice new values. Indeed, time is, if nothing else, the dimension for expanding possibilities otherwise constrained by space. Time permits the variation that space fails to accommodate. And that is what evolution is all about: evolution is the road of potential and opportunity!

Human beings with indefinite life spans and low fecundity will have the endless opportunity to achieve infinite possibilities and will, in their new condition, become the fountainhead of new human values!(?) In the world of mortal human beings, nations have dedicated themselves to securing life, liberty and the pursuit of happiness for their citizens. In a world of virtually immortal human beings, nations would dedicate themselves to other inalienable rights: to clean air, potable water, toxin-free earth, nourishing food, comfortable housing, competent health care, and equal opportunity.

But I cannot go too far down this utopian road without acknowledging some hazards. Of course, like mortals, if our descendents are to survive at all, they are well advised to avoid poverty, unemployment, crowding (on freeways, in hospitals, places of work and worship, public transportation, schools, theaters), social unrest, war, and other sources of human misery. Society must provide universal immunization of children, end deprivation, and prevent malnutrition. Seat belts should be used in all automobiles and public conveyances; safe sex must be universally available and practiced; and access to medical care should be provided for every human being. Beyond these prophylactic steps, we must clean up our act! We have turned the Earth into a toxic dump for pollutants of every variety—arsenic, dioxin, hexachlorobenzene, mercury, methoxychlor, PCBs, phthalic acid esters, radioactive waste—and must now remove the hazards or we will continue to suffer the consequences.

Then, if we implement a suitable plan for life, and death continues to evolve in its present direction, the potentially ageless human beings of the future will have a youthful life, brimming with vitality and vigor, freed from decrepitude, no longer weighed down by age or saddled to futility by senescence. The future will not be one of growing old forever or becoming Struldbruggs and falling apart eternally. Barring accident or crippling illness, individuals will remain at the peak of their prowess and intellectual abilities throughout their indefinite lifetimes. They will be as agile as gymnasts and as resilient as footballers, acquire languages with ease, be at the peak of their mathematical genius, and compose prose that reads as gracefully as poetry and poetry that reads as plainly as prose. Images, metaphors, symbols, and, yes, puns will flourish in the new age of beauty, humor, creativity, and communication. This is the future we must design and prepare for our descendents.

How many times will an immortal human being tell the same joke and find it funny; change jobs and still find a challenge; change spouses and still feel sexual excitement, companionship, and fulfillment? I take my cue from Arthur C. Clarke who, in *Greetings,* bears another vision of the future.

Clarke predicts "that boredom will replace war and hunger as the greatest enemy of mankind."[30] It is a world of mean-spirited human beings posturing and threatening, looking for weaknesses and seizing opportunities to assert superiority and acquire possessions. It is a world in which need has disappeared but has not been replaced by anything creative—a world in which "ultraintelligent" machines so completely run Earth by ineluctable laws of efficiency that there is nothing for human begins to do whatsoever. But Clarke then finds a crack in the cosmic boredom.

> *Need* is the operative word here. Perhaps 99 percent of all the men [*sic*] who have ever lived have known only need; they have been driven by necessity and have not been allowed the luxury of choice. In the future, this will no longer be true. It may be the greatest virtue of the ultraintelligent machine that it will force us to think about the purpose and meaning of human existence. It will compel us to make some far-reaching and perhaps painful decisions, just as thermonuclear weapons have made us face the realities of war and aggression, after five thousand years of pious jabber.[31]

In the end, Clarke's optimism prevails. He reminds the reader that "all normal children have an absorbing interest in the curiosity about the universe. Which if properly developed could keep them happy for as many centuries as they wish to live," and he concludes, "[o]ur prime goal will no longer be to discover but to understand and to enjoy."[32] Thus juvenilization or neotenization of human beings may very well be our salvation, extending to curiosity and the youthful drive "to understand and to enjoy," subverting boredom while saving the world from war.

Like Clark, I can imagine a world of gentle-spirited human beings pausing to ponder while probing and poetizing. I can imagine the superannuated young of the future who are also "superintelligent." I doubt if aggression would be the norm of communication in the world operated by such people. Life will no longer be played out as a zero-sum game in which one person's gain is another person's loss. Pride, envy, and cupidity will disappear and our prime goal, too, will become understanding and enjoying.

Appendix:
Different Forms of Life and Death

In the course of life's 3.8 billion–year history on Earth,[1] the first-order strange attractor recognizable as "alive" would have consisted of a material capable of repeatedly organizing structure around it. In extant living things, a material of this sort would be called a "hereditary material." Today, hereditary materials are generally one of two kinds of nucleic acid: RNA or DNA.

The ability to organize materials can be local. Typically, *replication*, or DNA-dependent DNA synthesis, occurs when one strand of DNA orders the sequence of nucleotides in a complementary strand of DNA. Likewise, *transcription*, or DNA-dependent RNA synthesis, occurs when one strand of DNA orders the sequence of nucleotides in a complementary strand of RNA. The ability to organize material can also be wide-ranging. Unique parts of extant nucleic acids called *genes* have this ability to organize material beyond their immediate vicinity, in theory, as far as entire organisms. The *genome* is a composite of chromosomal genes, but living things utilize genes present in episomes, plasmids, plastids, mitochondria, and possibly other structures as well.

The ability to organize life arises from its fractal construction. Material encodes information which encodes and unfolds continuously. This extension of organization taking place in life goes beyond the organizing activity of genes. Following transcription, *translation*, or RNA-dependent protein synthesis, occurs when strands of RNA direct the ordering of proteins' subunits known as *amino acids* in polypeptide chains, which, in turn, are folded and molded into active shapes under the auspices of proteins called *chaperones* and degraded under the impact of ubiquitin and proteasomes.

At least seven orders of strange attractors beyond hereditary material organize life, although one hardly imagines that this list is exhaustive. The first two orders of strange attractors are top fractals or major categories of living things: (I) noncellular life or viruses, (II) cellular life of both microscopic and macroscopic living things. Noncellular life is highly varied and found everywhere, but because of small size and "secretive" ways, the full extent of noncellular diversity and its ubiquity are only beginning to be appreciated by contemporary biologists. Cellular life is more familiar, and its diversity and pervasive distribution make it relatively well known.

The remaining orders of life's strange attractors are lower-order fractals or subcategories. The first two comprise noncompartmentalized cellular life, or prokaryotes (IIA), and compartmentalized cellular life, or eukaryotes (IIB). The prokaryotes, in turn, fall into two domains (sometimes called kingdoms), Bacteria and Archaea, while compartmentalized life consists of one domain (or kingdom), known as Eucarya.[2] Eukaryotes, however, fall into additional subcategories: simple (IIB1) and compound (IIB2), and the compound fall into modular or colonial (IIB2a), individual (IIB2b), and eusocial (IIB2c) categories corresponding roughly to tissue, organ, and superorganismic grades of life. The pattern seemingly formed by strange attractors traces a course of evolution through living things, but this pattern does not correspond to that typically assigned to the evolution of phyla or *Baupläne*.[3]

(I) NONCELLULAR LIFE (VIRUSES)

Viruses (including bacteriophage, also known as phage) turn up in two states: a transitory extracellular state and a residential or intracellular parasitic state. In transit, the infectious particle, or viron, consists of hereditary material wrapped in a capsid of protein and sometimes enclosed in a fatty envelope decorated by glycoprotein (carbohydrate-protein complexes). Glycoproteins allow the viron to identify prey and effect infection, and upon successfully entering a host cell, viral enzymes contained in the capsid or those quickly translated from the virus's hereditary material establish the parasite in the hapless cell. Having completed the transit successfully, viruses either sequester themselves within the host's genome (as proviruses) or take over and monopolize the host cell's metabolism for the manufacture of viral particles. Like other parasites, a virus's structure would seem maximally reduced, but the viron's capsid and envelope also seem highly differentiated and adapted for infecting new host cells, while once let loose within a cell, the products of the viral genome seem spectacularly adapted for reproduction.

Different viruses utilize RNA or DNA in both single- and double-stranded varieties as their extracellular hereditary material, although both nucleic acids are not found within the same capsid. Some viruses may have evolved from the reduction of cellular parasites whose hereditary material is passed on as DNA. On the other hand, viruses utilizing RNA in heredity are possibly evolved from components of an earlier "RNA world"[4] where they may have been instrumental in the evolution of genes and chromosomes and, hence, in the sequence of events leading to cellular life.[5] Retroviruses, indeed, are famous for utilizing RNA as a template for manufacturing DNA (through the action of their viral enzyme, reverse transcriptase) and achieving quite the opposite of transcription. In fact, cell biologists utilize viral reverse transcriptase to manufacture "copy" DNA from RNA (known as cDNA), and cells might have originally manufactured so-called pseudogenes on RNA templates with the help of viral reverse transcriptase.

Viruses evolve via point mutations that alter hereditary material and, consequently, change the viron's infectivity or the virus's ability to interact with cellular metabolism and reproduce in or escape from the parasitic state. Viruses can also become extinct, as attested to by the success of the World Health Organization's drive to eradicate small pox. Polio is next on the list of viruses targeted for extinction, but, so far, it has eluded eradication.

Many textbooks claim that viruses are immortal, inasmuch as virons in the transit state are presumably as stable as organic chemicals, while the hereditary material in viruses's parasitic state may be incorporated into the host cell's genome and preserved with the host cell. This claim for immortality is flawed, however, since immortality is hardly certain for organic compounds subject to oxidation, degradation, and denaturation, and the duration of the host cell's life is inevitably fragile. Furthermore, following the reproduction of viral components, many viruses kill their host cell in the process of leaving it. In effect, by destroying their parasitic habitat, these viruses destroy themselves. Pathologic viruses that kill their host commit suicide en masse, although a few may escape to infect new hosts.

(II) CELLULAR LIFE

Cells are extant representatives of ancient life that utilized DNA as its exclusive hereditary material. And cellular life is inherently composite. Cells combine different viral virtues in structures that presumably evolved through the formation of alliances among originally separate viral-like elements. Hereditary material is permanently incorporated into the DNA of so-called chromosomes and passed on intact to new cells via replication and cell division. Naked genes may, however, move among cells, and genes combined with viral-like DNA may move between and within cells. Wrapped up in a cell's envelope (membrane), or plasmalemma, the cell's cytosol carries out translation and complex metabolism comprised of metabolic pathways. The qualities of metabolic proteins are dictated by hereditary material but go on to prescribe growth, development, differentiation, and maintenance.

All existing varieties of cellular life rely exclusively on double-stranded DNA as their hereditary material, while RNA acts exclusively as the template for the cell's proteins as well as performing some enzymatic and regulatory roles (such as gene silencing). Changes in DNA, or *mutations*, can damage cells by disrupting RNA functions. Cells die from several causes, including deprivation of required resources or excess of toxic materials.

Many cells also have the capacity to die through their own devices. Different mechanisms of programmed cell death cause cells to commit suicide individually or collectively.

(IIA) NONCOMPARTMENTALIZED (PROKARYOTES) CELLULAR LIFE

Prokaryotes, frequently called *microbes*, encompass the Bacteria (including mitochondria and chloroplasts) and the Archaea, each representing a prokaryotic domain.[6] The cells are typically described as single and living in suspension, although they may also live in multicellular fibers (or even more complex solid structures) and in two-dimensional mats or biofilms whose properties are strikingly different from those of isolated cells, notably resistance to antibiotics. Replicated DNA is redistributed to new cells during cell division, or binary fission.

Prokaryotic cells may contain a rudimentary cytoskeleton of filamentous organelles bathed in cytosol. The cells' hereditary material (DNA) is *not* segregated to a specialized compartment within the cell, and the translation of transcribed RNA begins immediately without its translocation to a cytoplasmic compartment. Prokaryotic cells generally contain no membranous organelles, although parts of the plasmalemma may act as an organelle. For example, localized stretching of the cell's plasmalemma distributes replicated DNA to new cells during binary fission, and in blue-green bacteria, convoluted membranes function as photoreceptors. The cell's surface may also be decorated with locomotory "flagella" and sexual pili through which DNA moves between copulating cells.

Prokaryotes' DNA is double-stranded and circular, and a molecule containing the genome may be accompanied by plasmids or additional double-stranded, circular DNA molecules. Prokaryotes may evolve via mutations, especially in plasmids that accumulate in cells following their movement via various mechanisms: (1) through the cells' environment (known as *horizontal gene transfer*), (2) through sexual pili during prokaryotic copulation, (3) and with the help of viruses (known as *transduction*). The DNA moved is not necessarily that of closely related prokaryotes. Indeed, a large part of the biotech industry depends on the ability of bacterial cells to incorporate foreign genes. Regrettably, a large part of the current spate of drug resistance in pathological microbes is also a function of the uptake of foreign genes (conspicuously the notorious R plasmid).

The physiological conditions supporting maintenance and division vary vastly among prokaryotes. Some archaeans (hyperthermophiles living in bubbling hot springs and deep ocean vents) thrive in conditions that would immediately destroy many other forms of life. These prokaryotes would certainly deserve their reputation for immortality if resistance to adversity were the only criterion.

Because the products of binary fission are not easily distinguished from each other, prokaryotes are sometimes said to be immortal. Of the two cells formed by binary fission, however, one has the preponderance of old membrane, and this cell has a reduced ability to divide compared to the other cell.

Growth and replication are prerequisites for the maintenance of prokaryotic cells and must be sustained if prokaryotic life is to continue. Prokaryotic cells, thus, can die under a variety of circumstances: when experiencing deficiencies of resources; when exposed to excesses of unfavorable substances; when infected by phage, and so on. Prokaryotes disintegrate and disappear when physiological conditions do not prevail and the cells are unable to continue growing and dividing. What is more, some prokaryotes exhibit programmed death or regulated death rates via a "killer protein" and a cell-to-cell communication circuit involved in adjusting cell density in response to environmental clues.[7] In addition, members of the Bacillus/Lactobacillus division of Bacteria respond to unphysiological conditions by initiating encystment. Following a cell division, one of the cells encapsulates the other, dying while forming a virtually impenetrable protective cyst. The dormant endospore within the cyst is capable of surviving extreme conditions that could kill most cells. Thus, encapsulating bacteria would seem to exhibit a form of altruistic cell death, dying for the sake of potential survivors who might continue the line in the future, when better days prevail.

(IIB) COMPARTMENTALIZED OR NUCLEATED CELLULAR LIFE (EUKARYOTES)

Compartmentalized cells consist of two fundamental compartments: the nucleus, delineated by a nuclear envelope, and cytoplasm, delineated by a cell membrane or plasmalemma. The nucleus contains linearized, double-stranded DNA (as opposed to circular DNA), ending in monotonously repeating sequences (*telomeres*) and packaged in chromatin consisting of nuclear proteins and nucleosomes where DNA is wrapped around basic proteins known as *histones*. The cytoplasm consists of membranous organelles (the endoplasmic reticulum, golgi apparatus [also known as the dictyosome], lysosomes, etc.), mitochondria and plastids, such as chloroplasts (although plastids and mitochondria may also be absent), as well as a cytoskeleton of filamentous organelles and a suspending medium or cytosol. The nuclear envelope contains pores through which the nucleus and cytoplasm communicate. The plasmalemma may be fashioned into flagella (also called cilia when numerous) or undulipodia, short projections (microvilli), junctions (and junctional complexes) that link cells together or fasten cells to an extracellular substratum, and receptors through which cells transduce information (messages) from their environment.

Replication and transcription take place exclusively within the nucleus, while translation takes place exclusively within the cytoplasm. Thus, following synthesis and processing, mRNA is translocated to the cytosol before it is translated or destroyed. The links between nucleus and cytoplasm are complex, and complexity is compounded by the folding and modification of proteins, some of which are transferred back to the nucleus where they may act as transcription factors, receptors, or both, and bind with DNA thereby regulating the activity of genes.

Life's "nanomachines" of translation, known as *ribosomes*, are generally distinctly different in eukaryotes and prokaryotes, but there are exceptions. Ribosomes and their subunits, with their unique ribosomal RNA (rRNA), are identified by sedimentation constants (known eponymously as Svedberg [S] units). Bacteria have 70S ribosomes with a small subunit (30S) containing 16S rRNA, and a large subunit (50S) containing 23S rRNA, and 5S rRNA, whereas eukaryotes have 80S ribosomes with a small subunit (40S) containing 18S rRNA and a large subunit (60S) containing 28S rRNA, 5S RNA, and 5.8S rRNA. The exceptional eukaryotes are microsporans that have bacterial-type ribosomes, subunits, and rRNAs.

Cell division takes place following replication and the condensation of nuclear chromatin into chromosomes. DNA is distributed qualitatively and quantitatively equally to the new cells by a mitotic apparatus, whereupon the chromosomes decondense and DNA returns to chromatin.

Many genes originally present in mitochondria and plastids have moved permanently to the nucleus (via *lateral gene transfer*) and become parts of the host cell's chromatin. Other genes in the nucleus are thought to be largely responsible for development, differentiation, and maintenance of eukaryotic cells, while the remixing of genes through sexual reproduction is thought to be largely responsible for variation within species and hence evolution. Genes may also change by mutation, but the sorts of changes that give rise to evolution seem to result from changes specifically in genes that regulate the activity of other genes.[8] Evolutionary novelties may also result from genomic doubling, hybridization, and new combinations of genes resulting from errors in sexual reproduction.

Death is a constant companion of the eukaryotic cell. Cells within the same organism may kill one another, and cells may initiate their own destruction via programmed cell death (PCD). Some cells may enter mitotic (aka vegetative) senescence after a particular number of cell divisions, and cells may commit suicide through a morphological process known as *apoptosis*, frequently initiated by mitochondria. Other cells may harden, denature, or dissolve themselves en masse through a process known as *autophagia*.

The potential of eukaryotic genes for organizing complexity seems to have provided the substrate for the evolution of multicellular complexity in the bodies (thalli or soma) of so-called higher organisms. Sex cells (known as germ cells, gametes, eggs, and spermatozoa) also differentiate. And both body cells and sex cells die following differentiation.

(IIB1) SIMPLE COMPARTMENTALIZED CELLULAR LIFE (UNICELLULAR, PLASMODIAL, AND MULTICELLULAR)

Simple compartmentalized cellular life spills over into a host of organisms once classified as protozoa, algae, fungi, and other sundry groups. Today, their taxonomy remains unsettled, and simple compartmentalized organisms (euglenids, radiolarians, dinoflagellates, slime molds, ciliates, foraminiferans, "sporozoans," diatoms, and so on) are lumped with compound compartmentalized organisms (red, green, golden-brown, yellow-green, brown algae) in the Protoctista (eukaryotic micro-organisms and some of their descendents exclusive of animals, plants, and most fungi).[9] Fundamentally, simple compartmentalized cellular life consists of amoebae, or cells with cytoplasmic extensions known as pseudopods, and mastigophorans, or cells with whiplash flagella (also called undulopodia), but cells may alternate between forms and both types may also get together in small multicellular colonies or in fusion masses called *plasmodia*.

Small cells with one nucleus or a relatively small number of nuclei tend to have more rapid rates of cell division while large cells tend to become conspicuously multinucleate or united in multicellular organisms. Nuclei may have one copy of a genome, although even some small cells have multiple copies of their genome within nuclei (that is, achieve a high degree of polyploidy). Cells vary from free-living foragers to obligatory parasites (including the human pathogens *Entamoeba histolytica* and *Giardia intestinalis*) and from actively eating to encysted dormant stages. Photosynthesis is also common, and cells may lack a distinct oral apparatus, although heterotrophs may ingest (phagocytize) or engulf food particles through the action of cytoplasmic extensions known as *pseudopods* or drink up nutrients (via pinocytosis or endocytosis).

Pseudopods vary from lobular to elongate, flattened and fan-shaped to tubular, rounded to pointed (conical to filose), unbranched to branched, and from transient to stable. Taxonomy is sometimes based on the morphology of pseudopods or flagella, possibly the single most widespread organelle found among members of this group, although some naked and testate (shelled) amoeboid forms lack flagella entirely. Other organelles, including mitochondria, once thought to be diagnostic features are now considered "suspect," since their absence in parasitic forms may be due to loss in the course of evolutionary history rather than a failure to have evolved (or been acquired) in the first place.

Death, it would seem, is a universal feature of the colonial, plasmodial, and multicellular stages of simple compartmentalized cellular life. The cellular slime molds (Phyla Acrasea and Dictyostelida), for example, live freely as amoebae (myxamobae) until they aggregate (the pseudoplasmodium stage), form multicellular "slugs" that wander, settle, and produce stalked (more differentiated in the dictyostelids than the acrasids), multicellular aerial "fruiting bodies" (sorocarps) that release cells (sorophores, spores, or thick-walled cysts). Once germinated, these cells resume the amoeboid way of life without having, it would seem, any flagellated or sexual phase. The stalk and remainder of the fruiting body, however, are doomed to die.

Another simple compartmentalized cellular life form, the true or plasmodial slime molds (class Myxomycota), generally lives in litter on the forest floor as a massive, single cytoplasm (a plasmodium) containing many nuclei. Remarkably, these nuclei divide simultaneously and synchronously. The plasmodia also form stalked fruiting bodies, or sporophores, which release single cells, or spores, from apical sporangia, some of which cells, called myxamebas, go on to form swarm cells that conjugate (copulate). The product of this sexual union, or syngamy, then goes on to form a new plasmodium, while the remains of the sporophore die.

(IIB2) COMPOUND UNICELLULAR AND MULTICELLULAR COMPARTMENTALIZED LIFE

The difference between simple and compound compartmentalized cellular life is the degree of fractal expansion. Instead of equipotential genomes present in all the nuclei, in compound unicellular forms, somatic functions are relegated to a *vegetative nucleus* (in the trophozoite stage) and generative functions to a *micronucleus*. Moreover, instead of equipotential cells, *somatic* and *generative functions* may be relegated to different cells. These trends toward a division of labor seem to have continued in various modalities in compound forms leading to the intricate cortical structure and oral apparatus of complex unicellular forms and the extensive differentiation of tissues, body parts, organs, and organ systems among multicellular organisms.

Some compound unicellular forms, such as myxozoans, exhibit cellular junctions similar to those between some somatic (epithelial) cells of compound multicellular forms. Similarities extend to high affinities in their rRNA and protein sequences.[10] Indeed, "[p]hylogenetically, the combination of molecular and morphological data provides robust evidence in support of the inclusion of the myxozoans within the [multicellular] phylum Cnidaria,"[11] but whether similarities between compound multicellular and compound unicellular organisms are examples of evolution from a common ancestor or convergences among unrelated organisms remains uncertain.

Unicellular compound compartmentalized life is vulnerable to death, but frequently foregoes death in favor of sex. For example, in the well known ciliate, *Paramecium*, "as a general rule, isolate cultures are mortal and must eventually die out,"[12] but if the opportunity to swarm, encyst, and conjugate appears, which is to say, if the cells can undergo sexual union, life goes on, at least in a proportion of the successfully mated individuals. Seventy to eighty percent of the zygotes or exconjugants, formed following separation of the conjugants give rise to new lines that resume the cycle of rapid vegetative growth until they too get the urge and either conjugate or die out.[13]

The cells in multicellular compound forms do not resemble isolated cells so much as cells cooperating in a mutual enterprise. Small and large multicellular *colonies* consist of clonally derived cells (from a single originary cell) with allegedly identical nuclei. These colonies may be organized as solid masses (rods and spheres) or in sheets (sealed into tubes and balls) that resemble tissues, in the sense of broadly similar cells performing similar or related functions. Typically, more compound life forms develop when clones of cells remaining together following division form larger, cooperative units, such as organs and organ systems.

In relatively small colonies, groups of cells become isolated and differentiated as sexual cells or gametes. These cells are capable of fusing in fertilization and giving rise to new cells. Sex cells, thus, form a germ line, although prior to their differentiation as gametes, they may be identical with other cells in the colony. Somatic cells may show additional specializations, such as sensitivity to light or strength of motor activity. Death may occur in any member of a compound life form, but many organisms having this form replace parts continuously and show no signs of aging despite cellular loss.

In relatively large, compound multicellular forms, especially animals, the soma and germ lines tend to separate more completely and earlier in development than in small, colonial forms and plants. Somatic cells continue to share the same genome, but their differentiation becomes orders of magnitude greater than that achieved in other forms. When this highly differentiated soma fails to achieve immortality, its death is catastrophic, and the entire organism dies, including its germ cells.

(IIB2A) COMPOUND MODULAR AND SUPER-COLONIAL LIFE

Modular and super-colonial life is also fractal, consisting of self-similar parts, and is represented by compound cellular forms typically placed among the fungi, plants, and animals such as corals. Modules are clones or colonies, and, bound together with other similar units, comprise a compound super-colony of units functioning collectively.[14] Cells in modules and individual modules may be organized differently from each other and perform specialized functions.[15] For example, in free-floating super-colonies (such as the Portuguese man-of-war), different modules function primarily in floatation, defense, feeding, possibly navigation (homing), regulation, restoration, and sexual reproduction. In modular colonies, asexual reproduction may be limited to members capable of regulation and restoration, or individual units may break away and regenerate the whole. The super-colony, thus, may survive indefinitely as long as its rate of cellular turnover is equal to or greater than its rate of cell loss.

The differentiation of sex cells capable of fusing in fertilization and giving rise to new organisms creates a germ line, although prior to their differentiation, these cells are identical with other cells in the super-colony. Other cells in the super-colony, capable of differentiating in other directions, form the soma. In both modules and the super-colony generally, the life cycle involves a phase of development during which new members are added quantitatively and differentiated qualitatively. Thus, stages of development become an aspect of growth in modules and super-colonies. Likewise, death may occur at any stage during development. Again, however, cells comprising the tissues of modules and super-colonies may undergo turnover, and many modules are also replaceable. Germ and somatic cell lines having continuous replacement despite cellular loss may show no signs of aging or death.

(IIB2B) COMPOUND INDIVIDUAL LIFE: "ORGAN-ADDERS"

"Organ-adders,"[16] including us, are fractal organisms that acquire complexity not by adding modular units but by compounding tissues, organs, and organ systems in self-same units. Death of the body becomes conspicuous here if only because the separation of the body and germ occurs early, and the body is multiply differentiated compared to the germ.

Here then is the great enigma of differentiation: Why do some cells give up some functions to other cells? Or, in the extreme case of Weismann's divide (see chapter 3): Why do cells in the somatic line give up immortality, while other cells in the germ line retain it? The now-classical answer comes in two parts. First, according to biology's canon, all the cells in organ-adders share a common origin and thus a single genome. Second, cooperation, or *eusociality* pays off in the long run and "the cells of a multicellular individual are euso-, cial."[17] In the algorithm of compound, multicellular life, altruistic acts performed by cells of the somatic line are rewarded by the direct reproduction of other cells sharing the same genome. Hence, the somatic line passes its own genome along to the next generation even without having progeny.

.

But does differentiation depend on cellular relatedness? Indeed, how closely related are the cells of organ-adders? The notion of *devolution* suggests that various parts of a genome may have distinctly separate origins and, hence, not be closely related. Devolution accounts for the stages organ-adders pass through in their lifetimes as a function of somatic cell lineages having been derived from different ancient parents.[18] These lineages coexist in a single devolved individual as a result of each lineage emerging at a single stage and retiring in favor of another lineage at the advent of the next stage. In theory, only a small part of a cell's genome is active at any time, and that part is liable to be different in different cells at the same time and in the course of time. If organ-adders actually evolved by adding genomes as a result of hybridization of egg and sperm across species' lines (*illicit fusion*), and the genomic activity in organ-adders is determined by the sequential activation of unrelated genomes (that is, by *nomadic development*), the altruistic acts of somatic cells are that much harder to explain.

Alternatively, passing on genes may not be the be-all and end-all of life. Life may, instead, revolve around self-similar strong attractors. Generations are, after all, nothing more than temporal fractals building up upon each other, while death is merely spatial fractals collapsing onto each other. From this fractal point of view, one would expect the germ and somatic lines to be somewhat fluid, adaptive, and permeable.

(IIB2C) COMPOUND EUSOCIAL LIFE: SUPERORGANISMS

Superorganisms consist of compound societies in which the individual is subordinate to the group. Members of the group may be differentiated morphologically (structurally and functionally) and/or behaviorally and recognized by caste, age, and status, but individuals within these subgroups are virtually interchangeable. Superorganisms have evolved broadly among animals—ants, bees, wasps, and termites, naked mole rats, spotted hyenas, African wild dogs, herd ungulates, and some eusocial spiders.

Typically, eusocial organisms are divisible into germ and somatic lines. In the case of highly eusocial insects, the superorganism may contain a reproductive alpha female, the "queen," a few reproductive males, and modestly reproductive or nonreproductive beta females. "[T]he queen and the reproductive males are analogous to the germ line of multicellular organisms, and the nonreproductive individuals would be the soma of the superorganism."[19] The division of labor continues in the "soma"—workers doing different tasks at different times or individuals in separate castes performing duties as soldiers, and foragers, and so on.

The naked mole rat and hunting prides take a similar course, if to a less degree, but other eusocial mammals reverse the pattern: an alpha male attaches himself to a group of reproductive females and the division of labor in the female group is subordinate. Germ functions are generally periodic or seasonal and may be combined with somatic functions. For example, during the rutting season, the alpha male red deer in a herd may also function as the "sense organ" and "nervous system" for the herd, watching and listening while others forage. At the first signs of approaching danger, his signal triggers an escape reflex that sends the herd fleeing.

Superorganisms, or colonies of eusocial animals, illustrate the fluidity of germ and somatic lines and the division of labor in the soma. Instead of organs performing particular functions in the same organism, the individual is subordinated and performs particular functions in the group.

One may wonder to what degree human societies are superorganisms. Does a human city resemble a termite mound? Those who have speculated on this question emphasize the role of member-recognition in a superorganism as opposed to individuation in human societies. Eusocial insects will welcome members of the superorganisms and will not hesitate to attack nonmembers, but, at least in theory, we are more accepting. The Golden Rule, "Do unto others as you would have them do unto you," elevates individuals above membership in the group. "The essential points are that, in higher animals [us], social interactions within a group depend on individual recognition, and that an individual's behaviour towards another depends both on genetic relatedness, and on a memory of previous interactions with that individual."[20]

Human society does not, therefore, qualify as an unqualified superorganism. Indeed, efforts to wrench human society into a superorganismic vice have had to rely on extreme police powers and have been a potent cause of insurrection. In modern times, several totalitarian dictators have been unseated, although others remain in place. Human society seems to be moving away from compound eusocial organization of the sort represented by superorganisms and toward a flowering of compound individual societies.

Notes

PREFACE

1. de Grey, 2004c.

INTRODUCTION: DEATH THE MYSTERY

1. This is not to say that biologists are keen to change human beings. Indeed, "Because most life scientists have, along with everyone else, dismissed out of hand any thought of a possible fundamental reordering of the body, they are at a loss as to how to judge the import of human cloning, for example, a method through which the body could conceivably be reconfigured for the better" (Gins and Arakawa, 2002, xvii).

2. According to some Christian fundamentalists, death is not a flaw but the appropriate end of life for unconverted human beings. God will provide until Christ's return, but at the time of the apocalypse, or showdown in the valley of Armageddon, all God's "political and religious opponents [will] suffer plagues of boils, sores, locusts and frogs during the several years of tribulation that follow" (Moyers, 2005, 23).

3. Wowk, 2004, especially 136.

4. de Magalhães, 2004, 4. Later in the same essay, de Magalhães defines aging as "a sexually transmitted terminal disease that can be defined as a number of time-dependent changes in the body that lead to discomfort, pain, and eventually death" (48).

5. Epistle of Pavl: Apostle to the Romanes, The. *83kb*. Chapt. VI. *5kb*, 1996–2003. For additional theories of death see Comfort, 1979 and Medvedev, 1990.

6. Genesis 3:19. 1996–2003.

7. Vaughan, 1650, "My God! thou that didst dye for me," 1650.

8. Douglas, c1855, "For bonnie Annie Lawrie, I'd lay me down and dee."

9. Shakespeare, 1605, *King Lear.* iv. i. 38, [1994–1995].

10. Derrida, 1992, 15.

11. Derrida, 1992, 41.

12. Nagel, 1992, 1.

CHAPTER 1. EVOLUTION: DEATH'S UNIFYING PRINCIPLE

1. In their erudite *Chance, Development, and Aging*, Caleb Finch and Tom Kirkwood (2000) gravitate toward chance (another label for the unknown) and away from genes as the component of life that shapes death.

2. For overview, see Jazwinski, 1996.

3. Medawar and Medawar, 1983, 66–67.

4. In the words of the Reverend Thomas Malthus, "I only conclude that man is mortal because the invariable experience of all ages has proved the mortality of those materials of which his visible body is made" (Malthus, 1970, 129).

5. The entire title of the piece was *Essay on the Principle of Population as it affects the Future Improvement of Society, with Remarks on the Speculations of Mr. Godwin, M. Condorcet, and other Writers*. It preceded the revolution in actuarial science inspired in 1815 by Joshua Milne (1776–1851) in *Treatise on Annuities and Assurances on Lives and Survivorship*, and thus lacked the statistical finesse of later work. Malthus, 1970, 71 and 73.

6. Quoted from Flew, in Malthus, 1970, 24.

7. Malthus, 1970, 103.

8. Malthus, 1970, 250. The *Summary* was first published as an appendix to the third edition *Essay on the Principle of Population* but was then published separately in 1830.

9. Darwin, 1995, 40.

10. Letter from A. R. Wallace to A. Newton, 1887, in Darwin, 1995, 189.

11. Todes, 1989, 19.

12. Bonhoeffer et al., 2004, 1549; also see Sanjuán, Moya, and Elena, 2004; but see Michalakis and Roze, 2004, who conclude that recombination is antagonistic when the genes involve lower fitness individually.

13. For a recent variation on the theme of altruistic death see Fabrizio et al., 2004.

14. In order to avoid confounding data on juvenile periods, including the period of hatching and neonatal depression, these distributions begin after an initial period, one day for *Drosophila* and twelve years for *Homo sapiens*. The data on *Proales* begin at parthenogenetic hatching. (These data were gather by Dr. Bessie Noyes [Noyes, 1922] and are cited in Pearls, 1924.) All data quoted from Pearl, 1924, 336–77, table 112. Typically, the number of survivors is plotted as a log function and the drop off is much steeper as a consequence. Here the number is plotted as a linear function in order to facilitate comparisons with figures presented elsewhere in this book.

15. I should point out that each curve refers to hatched, born, or pupated organisms and not to their embryonic or larval forms. Had these been included, the curves would have fallen initially before entering their first level period. For other examples of survivorship curves, see Finch and Kirkwood, 2000.

16. Of course, if life span is determined in terms of the median, it would seem to expand outward, following the curve, but the maximum duration of a lifetime would not have changed. Also see Genes/Interventions database, 2004, at http//sageke. sciencemag.org/cgi/genesdb/ for an up-to-date list of genes affecting longevity in model organisms.

17. Lansing, 1952.

18. Finch, 1990, 485; see "Lansing Effect" on pp. 483–86.

19. Hekimi and Guarente, 2003.

20. Rose, 1998. Also see Charnov, 1991.

21. Gavrilov and Gavrilova, 1991, 90.

22. Kimmins and Sassone-Corsi, 2005, 583.

23. Yu et al., 2004; Cuervo et al., 2004.

24. Cohen et al., 2004, 391.

25. Nemoto and Finkel, 2004.

26. Kirkwood (1992, 11915) puts it this way in his first paper on the subject: "This 'disposable soma' theory predicts that aging is due to the accumulation of unrepaired somatic defects and the primary genetic control of longevity operates through selection to raise or lower the investment in basic cellular maintenance systems in relation to the level of environmental hazard." For subsequent versions see Kirkwood, 1999; Kirkwood and Austad, 2000. But also see Sozou and Seymour, 2004.

27. Rowe and Kahn, 1998, 29. The MacArthur Foundation Study of Successful Aging is a long-term, multidisciplinary research program of the MacArthur Foundation Research Network.

28. de Grey, 2004a.

29. Olshansky and Carnes, 2001, 69.

30. Hall, 2003, 9. Others (Kirkwood and Cremerm 1982) tap August Weismann as the parent of "cytogerontology."

31. Gavrilov and Gavrilova, 1991, 22.

32. Williams, 1957; Williams, 1966; but also see Bell and Koufopanou, 1986.

33. Provine, 1971; Provine, 1980.

34. Lolle et al., 2005.

35. "[T]he first generation of evolutionists attempted to explain the origin and distribution of life, using the ideas and information at their disposal." Bowler, 1996, 372.

36. Olby, 2000. According to Bowler (1996, 34) "Bateson struggled hard to establish genetics at Cambridge. He was eventually successful in the sense that he got a chair (professorship) in genetics in 1907, but he left soon afterwards for the privately funded John Innes Horticulatural Institutition because he still could not get enough research funding at Cambridge."

37. According to Keller (2002, 126), the term "gene" was coined by Wilhelm Johannsen "in large part in order to liberate the study of genetics from the taint of preformationism carried by the image of a hereditary unit *qua* organism."

38. The importance of one or another difference was left open, and one decided what to make of a difference based on experience and familiarity with data, much like one decides whether to carry an umbrella based on the weather forecaster's prediction of rain. Today, "significance" is routinely assigned to probabilities of differences between variances computed for data and those expected from random distributions. Differences less than 5 percent but more than 1 percent are said to be "significant," while those less than 1 percent are said to be "highly significant." In practice, the computation of statistical differences is left to ratios among variances and algorithms for assigning probabilities to those differences.

39. *Drosophila*, once known as the "vinegar fly" or "banana fly," is classified in the dipteran family Drosophilidae or pomace flies and is not a true fruit fly classified in the family Tephritidae. *Drosophila* has only a week ovipositor and lays eggs *on* decaying fruit rather than *in* fruit as a true fruit fly (see Carey, 2003, especially p. 10).

40. Pearl, Parker, and Gonzalez, 1923.

41. Arantes-Oliveira, Berman, and Keyon, 2003, 611; also see Houthoofd et al., 2004; Coffer, 2003.

42. For example, Hekimi and Guarente, 2003; Guarente, 2003; Kaeberlein et al., 2004 (yeast); Forbes et al., 2004; and Valenzuela et al., 2004 (flies); and Hasty et al., 2003 (mice).

43. Coles, 2004, personal communication.

44. Davenport, 2004b.

45. Genes/Interventions database, 2004.

46. Finch, 1990, 49.

47. Kipling et al., 2004, 1427.

48. For an alternative to the current model systems, see Roth et al., 2004.

49. de Grey, 2004a.

50. Allard, Lèbre, and Robine, 1998, 63.

51. Pearl, 1922, 158, 160. Pearl repeats the conclusion with a quip (p. 198) where he summarizes his own work on *Drosophila*: "The best insurance of longevity is beyond question a careful selection of one's parents and grandparents."

52. Pearl and DeWitt Pearl, 1979, 132. Elsewhere, Pearl (1946, 53) credits Oliver Wendell Holmes with first articulating the sentiment: "One of the most often quoted things that Oliver Wendell Holmes ever said was that if one is setting out to achieve

'three score years and twenty,' the first thing to be done, 'some years before birth, is to advertise for a couple of parents both belonging to long-lived families.' 'Especially,' said he, 'let the mother come of a race in which octogenarians and nonagenarians are very common phenomena.'"

53. Pearl and DeWitt Pearl, 1979, 145.

54. Herskind et al., 1996, 319.

55. Ljungquist et al., 1998.

56. Kannisto, 1991, 116.

57. Pearl, 1922, 26. But also see Austad, 1997.

58. Smith, 2004.

59. Smith, 2004.

60. Shaw, 1970.

61. Pardee, Peinking, and Krause, 2004, 1446.

62. Davis and Lowell, 2004.

63. Timiras, 1994.

64. Finch, 1990, 500–501.

65. Mair et al., 2003, 1732. For example, according to Shimokawa and Yoshikazu, 1994, 255, DR "suppresses the promotion step of chemically induced tumors."

66. McCarter et al., 1994, 162.

67. Nelson, 1994, 49.

68. Longo and Finch, 2003, 1342.

69. Cutler, 1981, 43.

70. Sapolsky, Krey, and McEwer, 1986.

71. Coffer, 2003, 1.

72. Pearl, 1922, 109.

73. Some thermodynamic theorists always had problems with this notion. James Clerk Maxwell (1831–1879), known best for his work on electromagnetic radiation and wave propagation, in particular, objected and asked if a "demon" might impose an anti-thermodynamic bias toward an increasingly less uniform temperature distribution by separating fast- and slow-moving particles. See Prigogine and Stengers, 1984, 239.

74. According to Prigogine and Stengers (1984, 175), developing, maintaining, and evolving biological systems present "an improbable physical state created and maintained by enzymes resembling Maxwell's demon, enzymes that maintain chemical differences in the system in the same way as the demon maintains temperature and pressure differences."

75. Jacob, 1973, 200–201.

76. Bergson, 1998, 242.

77. Bergson, 1998, 244.

78. Bergson, 1998, 245.

79. Bergson, 1998, 248.

80. Heidegger, 1977, 6, 12, 16.

81. Heidegger, 1977, 23.

82. Wolfram, 2002, 552, 554.

83. Jacob, 1973, 299.

84. Foucault, 1980, 139.

85. Tipler, 1994, 18.

86. The concept of sensitive periods arose from "imprinting," or the tendency of newly hatched chicks, ducks, and goslings to follow a moving object with little experience of it, but sensitive periods are also invoked to explain other forms of learning, for example the ready acquisition of language in human toddlers. See Hind, 1982.

87. Chiang, 1984.

88. Chiang, 1984, 77.

89. Pearson, 1897.

90. Pearson's data evidently come from "*Whitaker's Almanack*. On p. 357 of the edition for 1894 . . . [where one finds] a table dealing with the number of survivors left year by year out of a company of a million entering life together." Pearson further acknowledges "kind permission of the Council of the Royal Society from my memoir on 'Skew Variation.'" The table itself, however, acknowledges "Ogle: 1871–1880" (Pearson, 1897, facing page 26).

91. Pearson, 1897, 28, 39.

92. The number given is 175 in Pearl (1922, 95).

93. Pearl, 1922, 101.

94. Pearson, 1897, 31.

95. Pearson, 1897, 30.

96. Pearson, 1897, 31.

97. Gavrilov and Gavrilova, 1991, 22. The "recent data" cited is from Diamond, 1987.

98. Susser, 1981, 88.

99. Ventura et al., 2004.

100. In Denmark, between 1978 and 1992, "the risk of spontaneous abortion [for women grouped by age] would be: 12–19 years, 13.3%; 20–24, 11.1%; 25–29, 11.9%; 30–34, 15.0%; 35–39, 24.6%; 40–44, 51.0%; and 45 or more, 93.4%." (Nybo Andersen et al., 2000; Heffner, 2004. But also see Winter et al., 2002).

101. Sauer, Paulson, and Lobo, 1996.

102. Data from tables 36–41: Centers for Disease Control and Prevention, 2003.

CHAPTER 2. CHARTING DEATH'S EVOLUTION
AND LIFE'S EXTENSION

1. Erickson et al., 2004.

2. Finch (1990, 8) defines life span as the "total life duration of an individual organism, from its earliest developmental phase, whether it is derived from an egg or from a vegetatively propagating clone, to its death in the adult phase that ordinarily culminates its life cycle." I will use "life span" more generally for the interval between the earliest developmental phase and death at any stage of the lifecycle. "Lifetime" designates the interval between birth and death.

3. See Pimentel, 2004, on malnutrition. Further, an enormous number of citations affirm one or another hazard of modern life. One might find a gateway to this literature in Zinsser, 1934; Susser, 1981; Cadbury, 1997; Finch and Kirkwood, 2000; Mitman, Murphy, and Sellers, 2004; Wadman, 2004; Anon., 2005; Butler, 2005.

4. Eastern Europe is a conspicuous exception: see Kannisto et al., 1994; but also see Vaupel, Carey, and Kaare, 2003.

5. Boldsen and Paine, 1995, 32–33.

6. Allard, Lèbre, and Robine, 1998, 62; for additional data see Arias, 2002; Troyansky, 1989; McManners, 1981.

7. Olshansky, Carnes, and Cassel, 1990, 635.

8. According to Manton and Stallard (1996, B373), "The expectation of continuing U.S. mortality decreases is based on jointly evaluating temporal changes in the mean and variance of the empirical age at death distribution, the potential distribution of individual life spans, and cohort size changes."

9. Vaupel and Jeune, 1995, 109.

10. According to Robert Young (March 20, 2004) "Demographers like Jean-Marie Robine have shown that when comparing a population base (say, 70 million) to one ten times larger (say, 700 million), the expected age of the oldest individual, given a life expectancy of 75, is 114 for the smaller group and 115 for the larger group. Obviously, the statistical benefits of increasing the population base are very limited."

11. Finch and Crimmins, 2004.

12. A caption in the Drug Enforcement Administration (DEA) Museum (2004) reads, "In 1985, the number of people who admitted using cocaine on a routine basis increased from 4.2 million to 5.8 million, according to the Department of Health and Human Services' National Household Survey. Likewise, cocaine-related hospital emergencies continued to increase nationwide during 1985 and 1986. According to DAWN [Drug Abuse Warning Network] statistics, in 1985, cocaine-related hospital emergencies rose by 12 percent, from 23,500 to 26,300; and in 1986, they increased 110 percent, from 26,300 to 55,200. Between 1984 and 1987, cocaine incidents increased fourfold."

13. Olshansky and Carnes (2001, 123) tell us, "In developed countries where most people die beyond the age of sixty-five, medical miracle makers are the primary

reason why death rates are declining, life expectancy is rising, and survival rates for most lethal diseases are improving so dramatically."

14. For present purposes, the "environment" consists of all factors directly affecting longevity.

15. Kannisto et al., 1994, 794.

16. Gavrilov and Gavrilova, 1991, 134.

17. Ministry of Health, Labour and Welfare, at http:/www.mhlw.go.jp/english/database/db-hw/lifetb03/1.html.

18. See Endler, 1986; Valentine, 2004.

19. Laland, Odling-Smee, and Feldman, 2004, 609.

20. Chiang, 1984.

21. Chiang, 1984, 246.

22. Gavrilov and Gavrilova, 1991, 38.

23. I was greatly aided by Daniel Shostak, M.P.H., M.P.P., in adapting and using this model.

24. I ignore fertilized eggs and morulas simply because estimates of death at these stages are too speculative at present.

25. A life table is fundamentally a quantitative description of the age distribution at death in a cohort of organisms (born at the same time, typically a year).

26. I do not propose modeling a third possibility—a nonuniform expansion of nonconjoined stages, since, at the present level of resolution, such a model would be indistinguishable from uniform expansion.

27. I recognize that dietary restriction can extend gestation in rats and hamsters: See DePaolo, 1994.

28. Centers for Disease Control and Prevention, 2004.

29. Finch and Crimmins, 2004, 1736.

30. Finch and Crimmins, 2004, 1739.

CHAPTER 3. RETHINKING LIFECYCLES AND ARROWS

1. From the *Oxford English Dictionary*: "life cycle. Also life-cycle. [f. life sb. + cycle sb.]

1. Biol. The series of developments which an organism undergoes in the course of its progress from the egg to the adult state. Also, an account of these developments.
• 1873 *Monthly Microsc. Jrnl.* X. 57 Thus the entire life cycle of this form is seen.

• 1894 *Pop. Sci. Monthly* June 272 Each species has two generations in its life-cycle.
• 1967 M. E. Hale Biol. *Lichens* ii. 27 The life cycle of fungi is completed when the vegetative thallus produces fruiting bodies that contain spores."

2. Mojzsis et al., 1996. But see erratum: Eiler, Mojzsis, and Arrhenius, 1997; Fedo and Whitehouse, 2002; Sano et al., 1999; Dalton, 2004.

3. Schopf, 1999.

4. Senapathy, 1994.

5. For exceptions see Hanyu et al., 1986; RajBhandary and Söll, 1995; Watanabe and Osawa, 1995.

6. Doolittle, 1999.

7. Pace, 1997; but also see Rivera and Lake, 2004.

8. For more on life's forms see appendix; also Shostak, 2001.

9. Vreeland, Rosenzweig, and Powers, 2000.

10. Aguilaniu et al., 2003; Ackermann, Stearns, and Jenal, 2003; also see Crawford, 1981; Guarente, 2003.

11. Stewart et al., 2005; but see Woldringh, 2005.

12. Nyström, 2002.

13. Sex is defined here as the cycle of nuclear/cellular events in which the amount of hereditary material within a cell is doubled by fertilization and halved by the reduction resulting from meiotic divisions.

14. Abbott, 2005, 27.

15. Roenneberg et al., 2004.

16. From *Oxford English Dictionary*: "mitre . . . perh. with reference to the early form of the episcopal mitre, which had a vertical band bisecting the angle at the top." Hence, mitosis might refer to the bisecting of chromosomes during nuclear division. Coined by Walther Flemming in 1882 from the Hellenistic Greek root for thread to denote nuclear figures.

17. I chose to emphasize reduction rather than meiosis as such on the authority of August Weismann, who scooped his contemporaries by identifying reduction as the essential process taking place in the yet-to-be discovered meiosis.

18. Used here, "animal" and "plant" refer exclusively to multicellular organisms. "Protozoa" refers to unicellular and oligocellular organisms, while "fungi" and "algae" have uni-, oligo-, and multicellular varieties.

19. Regrettably, even pretenders to biological expertise often misrepresent life-cycles. For example, we are told "the idea that individuals are disposable once their reproductive role has been accomplished remains a cornerstone of modern theories of the evolution of aging," despite the other well-known fact that "Everything that is

known about life on earth indicates that under normal conditions, the vast majority of organisms die early in life." The incompatibility of these two "facts" seems to have escaped the authors' notice (Olshansky and Carnes, 2001, 58, 67).

20. Tong, Meagher, and Vollenhoven, 2002.

21. Remark attributed to L. Wolpert by Slack, 1983, 3.

22. Diamond, 1987 (see chapter 1 above).

23. Simpson et al., 1987.

24. Macklon et al., 2002.

25. Boklage, 2005. Also see Landy and Keith, 1998.

26. Tong, Meagher, and Vollenhoven, 2002, 142.

27. Centers for Disease Control and Prevention, 2003, table 19.

28. Lahdenper et al., 2004.

29. Skurk and Hauner, 2004.

30. Nemoto and Finkel, 2004, 152.

31. For a classical case of confusing the disease with the organism's effort to localize damage see Arrasate et al., 2004.

32. Gavrilov and Gavrilova, 1991, 165.

33. Ariès, 1981, 585.

34. From *Stedman's Medical Dictionary*, (1990, 444): "disease . . . 1. Morbus; illness, sickness: an interruption, cessation, or disorder of body functions, systems, or organs. 2. A morbid entity characterized usually by at least two of these criteria: recognized etiologic agent(s), identifiable group of signs and symptoms, or consistent anatomical alterations.

35. Pearl, who looked at the embryonic sources of organs and tissues involved in these classes of disease, concluded that the majority of these diseases over a lifetime arise in tissues of endodermal origin (54 to 55 percent in the United States and England) and a large part of the remainder arise in tissues of mesodermal origin (30 to 37 percent; leaving 9 to 14 percent to the ectoderm). After the age of forty in females and sixty in males, however, diseases in tissues of mesodermal origin predominant. Disease in tissues of endoderm origin occur with higher percentages in Sao Paulo which may reflect standards of public health and sanitation. (See Pearl, 1922, 140 ff.). Most of these diseases also occur in organs and tissues with high rates of cellular turnover.

36. Indeed, the normally brown color of feces is due in part to bile salts formed in the liver from degraded hemoglobin and poured into the intestine via the bile duct.

37. The term "natural death"seems to have been coined by Karl Pearson (1897). Pearl, 1922, 223.

38. Possibly the most extreme statement of this point of view is found in Klarsfeld and Revah (2004, 168): "The 'germ line [of yeast].' Created by a series of first buds, is potentially eternal, even if each individual yeast is incapable of budding more

than twenty or thirty times. In fact, a mother yeast would need to divide only twice in order for the population to increase exponentially, with each daughter in turn producing two buds, and so forth."

39. Weismann, 1882.

40. Especially Weismann, 1912.

41. For example, Lansing, 1952.

42. Bowler, 1988, 116.

43. Mayr, 1982. Actually, Weismann's terms for inherited particles fell out of use following the coinage of "genes." "Germplasm" was largely replaced by "chromosomes" and later by nuclear DNA, and Weismann's many terms for nuclear determinants of development were discredited long before the notion of mRNA. See Hubbard, 1982.

44. de Beer, 1951; Gould, 1977.

45. Ruse, 1999, 231.

46. Bell, 1988, 2.

47. Hyman, 1940, 177.

48. Bell, 1988, 43.

49. In other protozoans, such as radiolarians, however, "sexual processes are probable but unknown" (Hanson, 1977, 265). For the contrary view see Margulis, Schwartz and Dolan, 1994.

50. Different species accompany the meiotic divisions with one or more mitotic divisions followed by the disintegration of excess nuclei.

51. In some strains, cells are capable of recombining the products of meiosis without conjugation, a process known as automixis, autogamy or endomixis, suggestive of hermaphroditism or self fertilization. And many ciliates encyst before undergoing nuclear reorganization.

52. Finch, 1990.

53. For references see Noodén, 1988, table 12.2; Finch, 1990, table 4.2.

54. According to Benecke (2002, 26, caption fig. 4), "Ostensibly the oldest tree in the world, a Huon pine stands on top of Mount Reid in Tasmania. Scientists estimate the enormous tree, or tree system, to be at least 10,500 years of age—perhaps even 30,000 to 40,000." The Web site for Tasmanian tourism, however, posts the age of the oldest tree at 2000 years. Go to http://www.tourtasmania.com/ and follow links to flora and huon pine.

55. National Park Service, http://www.nps.gov/grba/.

56. Finch, 1990.

57. Kishi et al., 2003. Even the warm-water representative, zebra fish (*Danio rerio*), is sometimes said to show very gradual or sub-negligible senescence, although

the claim has to be doubted, since zebra fish die within months even with the best of care. (See Cailliet et al., 2001).

58. Anon., 1954. Finch (2004) is skeptical.

59. Henry 2003; *eNature*, 2003.

60. Congdon et al., 2003.

61. Nicholls, 2004.

62. See Finch, 1990, chapter 4, for additional citations.

63. Anderson and Apanius, 2003.

64. Holmes et al., 2003.

65. George et al., 1999, 578.

66. Charnov, 1991.

67. Gavrilov and Gavrilova, 1991, 274.

68. For a recent, short summary on longevity in trees, regrettably lacking citations, see Krajick, 2003.

69. See Comfort, 1979, 110. Hydra seems immortal, although it is also reported to die in 4 to 167 days (see Pearl, 1928, 16). I have reared individual hydras in separate dishes for as long as 2 years. When fed freshly hatched brine shrimp 3, 5, and 7 days a week, the animals' cell density and size remained in a steady state apparently due to the release of excess cells through budding (Shostak, 1968; Shostak et al., 1968).

70. Bell, 1988, 131.

71. Kolata, 1998, 236–37.

72. Bell, 1988, 84.

73. Couzin, 2004.

74. Slowly dividing basal spermatogonia are self-renewing stem cells, dividing asymmetrically and producing precursor cells at the same time as new stem cells (see chapter 4). The precursor cells are not self-renewing but divide more rapidly and abundantly.

75. Eggs at earlier stages can fuse with spermatozoa but do not develop normally (See Baker, 1982).

76. Johnson et al., 2004.

77. Wise, Krajnak, and Kashon, 1996.

78. Perez et al., 2000. What is more, mice accumulating errors in mitochondrial DNA age precociously (see Trifunovic et al., 2004) and nuclear-encoded mitochondrial genes are downregulated in aging human brains (see Lu et al., 2004).

79. According to the *Courier Mail*April 26, 2004, reporting on comments by Drs. Natalia Gavrilova and Anna Clark attending a longevity conference in Sidney, Australia.

80. Bowler, 1988, 116.

CHAPTER 4. KEEPING LIFE AFLOAT

1. Williamson, 1992.

2. Saunders and Fallon, 1966.

3. Vyff, 2004, 224.

4. Courtillot, 1999.

5. Comins, 1993.

6. Shostak, Seth, 1998, 61.

7. Prigogine and Stengers, 1984, 163.

8. Prigogine and Stengers, 1984, 176.

9. Camazine et al., 2001, 26.

10. Lewin, 1992.

11. Boulter, 2002, 60.

12. Ho, 1998.

13. Yorke and Li, 1975.

14. Gleick, 1987, 69.

15. Kauffman, 1993; Kauffman, 1995.

16. Maturana and Varela, 1998, 46–47.

17. Gleick, 1987, 266–67.

18. Berlekamp, Conway, and Gay, 1982, 825.

19. Lorenz, 1993.

20. Conway Morris, 2003, 127.

21. Prigogine and Stengers, 1984, 167. Exactly what is intended here by "chaos" may seem vague. The authors have in mind nonequilibrium turbulent chaos and not equilibrium thermal chaos as they explain, later, on the same page: "The interplays among the frequencies produce possibilities of large fluctuations; the 'region' in the bifurcation diagram defined by such values of the parameters is often called 'chaotic.'"

22. The beautiful wave patterns captured by the Belousov-Zhabotinsky reaction in spreading cerium are probably the most familiar examples of the creative potential of far-from equilibrium thermodynamics operating in the nonliving world.

23. Brasier and Antcliffe, 2004; Narbonne, 2004.

24. Raup, 1991, 51.

25. Pearson (1897, 11) describes his "generalization of the theory of Chance . . . [as a] modern conception of the dance of Death."

26. Finch and Crimmins, 2004.

27. Exceptions are gamblers with trained memory playing black jack who turn the odds in their favor by memorizing cards already played and therefore not available for drawing off the top of the deck.

28. According to new life expectancy statistics for the United States, Americans are living longer: today's median is 77.6 years (females, 80.1 years; males, 74.8 years; delta, 5.3 years) (Dooren, 2005; AP, 2005; Mestel, 2005). Also see National Center for Health Statistics: http://www.cdc.gov/nchs 02/28/05 21:10.

29. *The Wall Street Journal Online*, March 10, 2005 asks, "Are Americans healthier now than they were in 1983? A recent Harris Interactive poll measuring the number of U.S. adults who smoke, wear seatbelts and are overweight found mixed results.

"Charting life expectancy: See a chart of the Harris poll data, tracking trends in smoking, obesity, and seat-belt use from 1983–2005.

"The number of Americans who smoke has declined from 30% in 1983, to an estimated 19% in 2005, according to the poll. And the survey indicates the number of adults who use seatbelts has soared, from 19% in 1983, to 86% in 2005."

30. During the 1990s, the population of Singapore rose from 1415×10^3 men and 1381×10^3 women to 1630×10^3 men and 1633×10^3 women (World Health organization Mortality Database, 2004).

31. Finch and Crimmins, 2004.

32. For the greater part of the life expectancy curve (derived from mortality data), the likelihood of death accelerates by nearly one (about 0.8) year per year per year, so that one year olds in 2000 are likely to live another 70 years, while, in 2069, they are likely to live only 14 years.

33. Gavrilov and Gavrilova, 1991, 80.

34. Gavrilov and Gavrilova, 2002; Gavrilov and Gavrilova, 2005.

35. Young, April 16, 2005.

36. Coles, March 12, 2005.

37. Allard, Lèbre, and Robine, 1998.

38. Carey, 2003, 7.

39. Gavrilov and Gavrilova, 1991, 129.

40. Gavrilov and Gavrilova, 1991, 149.

41. Cells and organs are only supplied by transfusion and transplantation in dire emergencies.

42. Pikarsky et al., 2004; Balkwill and Coussens, 2004.

CHAPTER 5. PUTTING CELLS IN THE PICTURE

1. For example, Maynard Smith, 1962.

2. Schwann, 1955.

3. Benecke, 2002, 21; Spalding et al. (2005) demonstrate via retrospective birth dating that non-epithelium of the jejunum and skeletal muscle of the rib cage turned

over in fifteen to sixteen years while occipital neurons of the human cortex are as old as the individual.

4. Laird, Tyler, and Barton, 1965, 244.

5. Mintz and Illmensee, 1975; Tarkowski, 1975.

6. If Evans and Kaufman (1981) are correct, ICM cells may yet form trophecto-derm.

7. See Shostak, 2002, 90–95.

8. Jones and Takai, 2001.

9. Labosky, Barlow, and Hogan, 1994; Allegrucci et al., 2005; Wobus and Boheler, 2005.

10. Shostak, 2002.

11. Höstadius, 1950; LeDourin, 1982; Morrison et al., 1999.

12. See Moore and Persaud, 2002.

13. Robin et al., 2003.

14. Boggs, Saxe, and Boggs, 1984.

15. According to Raff (2003, 3), these terms are "usually used interchangeably, but some use progenitor cell to refer to a cell with greater developmental potential than a precursor cell."

16. Leblond and Walker, 1956.

17. Slack, 2000.

18. Globerson, 1999; Sudo et al., 2000; Kim, Moon, and Spangrude, 2003.

19. Gage, 2000; Zhao et al., 2003.

20. Johnson et al., 2004; Hoyer, 2004; Spradling, 2004.

21. Grounds et al., 2002.

22. For information on hepatocytes, see Rhim et al., 1994; for pancreatic islets, see Dor et al., 2004.

23. Lincoln and Short, 1980.

24. Alonso and Fuchs, 2003.

25. Tumbar et al., 2004; Turksen, 2004.

26. Marshman, Booth, and Potter, 2002, 93. Also see Martin, Kirkwood, and Potten, 1998.

27. Schofield, 1978.

28. Orwig, Ryu et al., 2002; Orwig, Shinohara et al., 2002; Ryu et al., 2003.

29. Schultz, 1996.

30. Doetsch, 2003; Rice et al., 2003.

31. Clermont, 1962; Brawley and Matunis, 2004.

32. Chen, Astler, and Harrison, 2003.

33. Renaulta et al., 2002.

34. Beltrami et al., 2003; Torella et al., 2004; Ohinata et al., 2005.

35. Müller-Sieburg et al., 2002.

36. Seipel, Yanze, and Schmid, 2004.

37. The *Stem Cells Book* sponsored by the National Institutes of Health holds out hope for transdifferentiation under the aegis of plasticity. For a different slant on dedifferentiation and transdifferentiation, see Weissman, Anderson, and Gage, 2001.

38. Hayflick 1996, 2003, and elsewhere.

39. Kipling et al., 2004, 1428.

40. Stewart et al., 2005; Leslie, 2005.

41. Guarente, 2003. The notion of cells being born with a limited capacity for regeneration and running out of their regenerative capacity during a lifetime seems to have been anticipated by the novelist, Jack London early in the twentieth century (see Oriard, 2000).

42. Espejel et al., 2004.

43. Klapper, Heidorn et al., 1998; Klapper, Kuhn et al., 1998.

44. Leslie, 2004.

45. Bodnar et al., 1998.

46. de Lange, 1998.

47. Sharpless and DePinho, 2004.

48. Rudolph et al., 1999.

49. Hall, 2003, 82.

50. Hall, 2003, 199.

51. Davenport, 2004a. See also Driver, 2004.

52. Sharpless and DePinho, 2004.

53. Bertsch and Marks, 1974.

54. Wilson, 1896.

55. Wilson, 1925.

56. Macfarlane Burnet, 1969, 37.

57. Howard and Pelc, 1951.

58. Laird, 1965.

59. Lajtha, 1979, 23.

60. Robert, 2004, 1006.

61. van der Koory and Weiss, 2000.

62. Williamson, 1992. Also see Shostak, 2002, 153–54.

63. Echinoderms, see Davidson, 1989. Amphibians, see Klein, 1987; Jacobson, 1982, 1983. Fish, see Kimmel and Warga, 1986.

64. Typically thought to be at the eight- or sixteen-cell stage (Johnson, 1985), asymmetric division may also occur at the first division accompanying rotation and compaction.

65. Weissman, 2000, 1443.

66. Blau, Brazelton, and Weimann, 2001, 833.

67. Potten, 1979, 282.

68. Potten and Loeffler, 1990, 1009.

69. Morrison et al., 1999.

70. Robert, 2004, 1007.

71. Tumbar et al., 2004.

72. Blau, Brazelton, and Weimann, 2001, 832.

73. Finch and Kirkwood, 2000, 137.

74. Schiffmann, 1997.

75. Marshman, Booth, and Potten, 2002.

76. Potten, 1979.

77. Cairns, Overgaugh, and Millers, 1988; Marshman, Booth, and Potten, 2002.

78. Raff, 2003.

79. Holtzer, 1979.

80. Rao and Mattson, 2001.

81. Stevens, 1960; Stevens, 1975; Damjanov, Damjanov, and Solter, 1987; Martin, 1981; Rudnicki and McBurney, 1987; Smith, 2001.

82. Evans and Kaufman, 1981.

83. Weissman, 2000, 1443.

84. Geiger et al., 1998; van der Koory and Weiss, 2000.

85. Jackson et al., 2001; Castro et al., 2002; Mezey et al., 2003; Shostak, 2006.

86. Allsopp and Weissman, 2002. Kanatsu-Shinohara et al., 2005, find telomeric shortening in spermatogonia in vitro.

87. See Mattson and Van Zant, 2002, vii.

88. Gavrilov and Gavrilova, 1991.

89. "Embryonic stem cells have the unique ability to form all adult cell types. Harnessing this potential may provide a source of cells to replace those that are lost or impaired as a result of disease." Cowan et al., 2004.

90. For example, Reya and Clevers, 2005.

91. But also see Bukovsky, Svetlikova, and Caudler, 2005. In this case, a synthetic growth factor, phenol red (PhR), seems to promote the differentiation of ovarian surface cells in vitro to oocytes and granulose (follicle) cells.

92. Blau, Brazelton, and Weimann, 2001, 832.

93. Abkowitz et al., 2002.

94. Winitsky et al., 2005.

CHAPTER 6. NEOTENY AND LONGEVITY

1. From *Oxford English Dictionary*, second edition, "neoteny . . . [ad. G. *neotenie* (J. C. E. Kollman, 1884, in *Verh. Naturf. Ges.* Basel VII. 391), f. Gr. *neoj* young + *teinein* to extend.]"

2. Pelicci, 2004. See also Sharpless and DePinho, 2004.

3. Gould, 2000; Godfrey and Sutherland, 1996; but also see Rice, 1997.

4. Noto and Endoh, 2004.

5. Finch, 1990.

6. De Beer, 1951, and Berrill, 1961; Gould, 1977.

7. Holden, 2004.

8. Finch, 1990, 628.

9. Safi et al., 2004.

10. Bowler, 1996, 150. Bowler acknowledges Roberta Jane Beeson's Oregon State University Ph.D. thesis, *Bridging the Gap: The Problem of Vertebrate Ancestry, 1859–1875*, as his source and credits her translations of original work. Kowalevsky's evidence is presented in several papers published from 1866 to 1871 on the comparative embryology of the lancet, *Amphioxus lanceolatus* (*Branchiostoma*), and a "simple" ascidian.

11. Garstang, 1985.

12. Hyman 1959. See also Knoll and Carroll, 1999; Rieger, Ladurner, and Hobmayer, 2005.

13. Roche, 1995, 13.

14. McDowell, 2003, 6.

15. Finch, 1990, 629.

16. Perls, Alpert, and Freets, 1997; Perls and Silver, 2000.

17. Turner, 2002. The quotation is attributed to the evolutionary biologist Marc Tatar.

18. Washburn, 1981, 23.

19. Dean, 1987, 213.

20. Bromage and Dean, 1985, 526.

21. Penin, Berge, and Baylac, 2002, 50.

22. Duque-Parra, 2003.

23. As exemplified by the structural organization of the bilingual brain as a function of the age of language acquisition. See Mechelli et al., 2004.

24. Allard, Lèbre, and Robine, 1998, 72.

25. Coqueugniot et al., 2004.

26. Susser, 1981, 81.

27. Gavrilov and Gavrilova, 1991, 115, 163.

28. Cutler, 1981, 36 and 47.

29. Mathe, 1997.

30. Eaton et al., 2004.

31. Franceschi, Bonafè, and Valensin, 2002, 1718.

32. Pawelec, Adibzadeh et al., 1997.

33. Szabo et al., 1998.

34. Toichi et al., 1997.

35. Dominguez-Gerpe and Rey-Mendez, 1998.

36. Chakravarti and Abraham, 1999.

37. Franceschi et al., 1996.

38. Bodey et al., 1999.

39. Ye and Kirschner, 2002.

40. Takeoka et al., 1996; Aspinall and Andrew, 2000; Andrew and Aspinall, 2002; Brelinska, 2003.

41. Lau and Spain, 2000.

42. Thoman, 1997.

43. Bar-Dayana et al., 1999.

44. Li et al., 2003.

45. Ortman et al., 2002.

46. Toichi et al., 1997.

47. Pawelec, Remarque, et al., 1998.

48. Lloberas and Celada, 2002.

49. Pawelec, Barnett, et al., 2002.

50. De Haan and van Zant, 1999.

51. Migliaccio et al., 1999.

52. Wagers et al., 2002.

53. Harris et al., 2004.

54. Hübner et al., 2003; Toyooka et al., 2003; Clark et al., 2004; Surani, 2004.

55. Geijsen et al., 2004, 149.

56. Howell et al., 2003.

57. For example, Blau, Brazelton, and Weimann, 2001.

58. Extavour, 2004; Donnell et al., 2004.

59. Conklin, 1986; Wilson, 1904a, b; Costello and Henley, 1976; Astrow, Holton, and Weisblat, 1987; Extavour and Akam, 2003.

60. Sulston et al., 1983; also see Strome and Wood, 1983.

61. Mahowald et al., 1979; Mahowald, 1983.

62. Eddy and Hahnel, 1983.

63. Dennis Smith, 1964.

64. Blackler and Fischbert, 1961; Blackler, 1962; Blackler and Gecking, 1972.

65. Tiedemann, 1975; Nieuwkoop and Sutasurya, 1979; but see Mahowald, 1977.

66. Willier, 1937; Reynaud, 1969; also see Reynaud, 1976.

67. Gardner, 1978.

68. Mintz and Illmensee, 1975.

69. Eddy, 1975; Eddy et al., 1981.

70. Shamblott et al., 1998.

71. Hübner et al., 2003, 1251.

72. Clark and Eddy, 1975.

73. Approximation estimated for 2002 data by *Global Population Profile*, 2002. For current population survey see http://www.census.gov./population/www/socdemo/fertility/sa04.pdf. For crude birth rates in the United States see http://www.cdc.gov/nchs/data/hus/tables/2003/03hus003.pdf.

74. Fehr, Jokisch, and Kotlikoff, 2004.

75. *Global Population Profile*, 2002, 3.

76. Social Security Trustees' Annual Report, 2004, 6.

77. Demeny, 1968.

78. Butler, 2004.

79. Caldwell, 1976, 323.

80. Caldwell, 1976, 358.

81. Smith et al., 2003.

82. Rose, 1998. Also see Charnov, 1991.

83. Lahdenper et al., 2004, 178.

84. Hawkes, 129, 2004.

85. Aitken, Koopman, and Lewis, 2004.

86. Plas et al., 2000, 543.

87. Carlsen et al., 1992; Auger et al., 1995; Sherins, 1995; Irvine et al., 1996.

88. Nesse and Williams, 1996; Kirkwood, 1999.

89. Charnov, 1991, 1137.

90. Tatar, Bartke, and Antebi, 2003.

91. Laland, Odling-Smee, and Feldman, 2004. But also see Keller, 2003.

92. Lovelock, 1990; Lovelock, 1996.

93. Kamshilov, 1976; Lapo, 1982.

AFTERWORD

1. Booke of Psalmes. PSAL.XC 10: "The dayes of our yeres are threescore yeeres and ten, and if by reason of strength they be fourscore yeeres, yet is their strength labour and sorrow: for it is soone cut off, and we flie away."

2. Shakespeare tells us in *Macbeth* that one is old at threescore and ten, but in the *Merry Wives*, one may yet live "foure-score yeeres and vpward."

3. Malthus, 1970, 126. Malthus reiterates his claim that "no decided difference has been observed in the duration of human life from the operation of intellect, and mortality of man on earth seems to be as completely established . . . [as part of his criticism of the] conjectures of Mr Goodwin and Mr Condorcet concerning the indefinite prolongation of human life" (157).

4. Malthus, 1970, 127.

5. Born Feb. 14/17, 1766, Rookery, near Dorking, Surrey, England; died Dec. 23, 1834, St. Catherine, near Bath, Somerset.

6. Malthus, 1970, 258–61.

7. According to Peter W. Frank (2004), the seventeenth century boasted some super-duper centenarians including, "Thomas Parr, who died in November 1635 at the alleged age of 152 years; Henry Jenkins, who died in December 1670 at the alleged age of 169 years; and Catherine, countess of Desmond, who died in 1604 at the alleged age of 140 years." But see Austad, 1997.

8. Frank (2004) also lists "eight individuals for whom records substantiate the fact that each had lived more than 108 years. . . . Six of the eight were more than 110 years old at death. The oldest was Pierre Joubert, who was born July 15, 1701, and died November 16, 1814, aged 113 years and 124 days."

9. Malthus, 1970, 129.

10. Allard, Lèbre, and Robine, 1998.

11. Malthus, 1970, 148.

12. Cutler, 1981, 36.

13. Freitas, 2004, 87.

14. de Grey, Feb. 15, 2004.

15. de Grey, May 3, 2004

16. Stearns, 1992, 127.

17. These data are for the first complete death rate tables by year of birth (cohort) in the online Human Mortality Database at www.mortality.org.

18. For example, in Sweden the median life expectancy for women was 68 years and for men was 65 years in 1939. These data are from life expectancy tables by year in the Human Mortality Database (pages 1–2).

19. See Pear, 2004.

20. Social Security Trustees' Annual Report, 2004, 3.

21. Social Security Trustees' Annual Report, 2004, 16.

22. Social Security Trustees' Annual Report, 2004, 3.

23. Blackford, 2004.

24. Geddes, 2004, 243.

25. Fukuyama, 2002, 173.

26. Fukuyama, 1992, 150.

27. Fukuyama, 1992, 329.

28. Kass virtually equates "living" with the "perishable." See Kass and Wilson, 1998, 19. Also Kass, 1983.

29. Gins and Arakawa, xviii–xix, 2002.

30. Clarke, 1999, 273.

31. Clarke, 1999, 271.

32. Clarke, 1999, 274–75.

APPENDIX

1. Furnes et al., 2004.

2. Woese and Fox, 1977; Woese et al., 1990.

3. See Shostak, 1999

4. Gilbert, 1986; Joyce, 1994; Joyce and Orgel, 1993; Woese and Pace, 1993.

5. Gilbert, 1985; Gilbert, 1987; Gilbert, 1992; Gilbert et al., 1995, 1997.

6. Woese, 1987; Woese, 1991; Woese, 1998.

7. According to You et al., 2004, 870, "Similar to multicellular organisms, bacteria possess sophisticated suicide machinery that is triggered by stress and starvation or by 'addiction modules' during post-segregational killing."

8. Shapiro et al., 2004; Shubin and Dahn, 2004.

9. Margulis, Corliss et al., 1990.

10. Smothers et al., 1994; Schlegel et al., 1996; Vossbrinck et al., 1987.

11. Siddal et al., 1995, 966.

12. Bell, 1988, 35.

13. Margulis and Sagan, 1997.

14. For example, see Hageman, 2003.

15. These definitions of module and super-colony are intended to accommodate the concept of fractals and, hence, are different from other definitions (see Maynard Smith and Szathmáry, 1999).

16. I borrowed the term "organ-adder" from Pearl (1946, 17) where he uses it specifically for the brain of *Homo sapiens*. Later, in the same volume, Pearl uses "environment-adder" for the far reaching effects of sociability on the environment.

17. Maynard Smith and Szathmáry, 1999, 126. These authors are relying on the concept of eusociality developed by Paul Sherman, Laurent Keller, and Nicolas Perrin.

18. "Illicit fusion is hybridization among unrelated species that does not produce a chimera so much as distinct stages of development. . . . Nomadic development is the sequential development of stages in a life history arising from the activities of separate genomes having disparate origins" (Shostak, 2002, 153. Also see Shostak, 1999).

19. Maynard Smith and Szathmáry, 1999, 126.

20. Maynard Smith and Szathmáry, 1999, 137.

Glossary

AC: adult cache: cells in adult expanding tissue.

actual or **functional stem cells**: the traditional stem cells of steady-state tissues.

AIDS: acquired immune deficiency syndromes.

allele: a particular gene or locus on a chromosome.

allometry: the correlation of growth of a part to the growth of a whole organism.

APCs: antigen presenting cells.

aptness: fitness; suitable integration of structure and function.

AR: adult reserve: cells in adult static and steady-state tissue; cognate of stem cell.

Archaea: one of the three domains of cellular life assigned originally on the basis of ribosomal RNAs; see prokaryote.

autopoiesis: immanent, self-regulating, or emergent abilities of organisms for development and maintenance.

Bacteria: one of the three domains of cellular life, including unicellular bacteria, biofilms, and blue-green bacteria, assigned currently on the basis of ribosomal RNAs; see prokaryote.

bell-shaped curve: see normal distribution.

bFGF: basic fibroblast growth factor.

blastocyst: early stage in development of mammals following cleavage and prior to implantation; aka preembryo.

blastomeres: cells formed by cleavage of the zygote and by further division of blastomeres prior to cell movement, rearrangement, and embryogenesis.

BM: bone marrow.

cache cells: differentiated chief cells of expanding tissues that retain capacity for proliferation.

***Caenorhabditis elegans (C. elegans)*:** a species of roundworms considered a model system for various kinds of cell and integrative research including aging.

chromatin: combination of DNA, and proteins within nucleus; strands of DNA wound around histones in nucleosomes.

chromosomes: [colored bodies] thread- or rod-like bodies that split and are equally distributed to the two cells formed in eukaryotes during cell division; linear nuclear bearers of genes and other DNA such as "junk" DNA and telomeres; also circular DNA of prokaryotes.

cleavage: division of the zygote and blastomeres.

cohort: all the members of a group born at the same time (typically a year for human beings; a day for fruit flies).

conjugation: sexual coupling of protozoans during which nuclei are exchanged and a new generation of exconjugants is spawned.

Darwinian evolution, theory of, or Darwinism: accounts for differences among species through the accumulation of small, quantitative variation; gradual change among species due to the accumulation of small, hereditary differences; differential rates of breeding among organisms with small, hereditary differences.

dedifferentiate: the loss of a cell's typical identity, typically associated with a change in a cell's determination.

deepithelialize: when an epithelial tissue gives rise to freely moving, independent cells.

determination: loosely, an irreversible phase in the chain of events leading to differentiation.

differentiate: a cell's acquisition of its final morphology; the appearance and accumulation of a cell's distinctive cellular content, typically intermediate filaments.

diploid: eukaryotic cell containing dual, homologous nuclear chromosomes.

DNA: deoxyribonucleic acid; genetic material of cellular life.

Drosophila melanogaster (Drosophila): a species of flies usually called fruit flies and frequently considered a model system for aging research.

EC: embryonic carcinoma.

EG: embryonic germ; pluripotent cells derived from germinal ridge of fetus and raised in tissue culture.

embryo transfer: see in vitro fertilization.

endopterygote (holometabolous): insects with complete metamorphosis in pupa; insect develops wings during metamorphosis in pupa.

environment: everything that impacts on an organism's ability to live and reproduce, from its abode, availability of resources, and the weather.

epiblast: one (unilaminar) or two (bilaminar) cell layers (also known as the embryonal plate) formed in the blastocyst from the inner cell mass and giving rise to the embryonic germ layers at gastrulation.

epigenetic: the host of controls, including gene silencing through DNA methylation, with their greatest roles in modulating patterns of phenotypic determination, non-Mendelian hereditary effects, and cytoplasmic influences on traits.

epistasis: interactions among genes, usually promoting some aspect of fitness in the phenotype.

EPL: early pregnancy loss.

ES: embryonic stem; pluripotent cells derived from epiblast of embryo and raised in tissue culture.

Escherichia coli (E. coli): model bacterium used in biotech industry; coliform; Gram-negative; enteric.

eukaryote: one of the three domains of cellular life; unicellular or multicellular organism whose cells are compartmentalized, containing membranous, and non-membranous organelles, a cytoskeleton, cytosol, and one or more nuclei bearing hereditary material in the form of nuclear genes; currently assigned on the basis of ribosomal RNAs.

exopterygote (hemimetabolous): insects molting through several larval stages without a pupa; insects developing wings during larval stages.

FACS: fluorescence-activated cell sorting.

founder cells: embryonic blast or stem cells in *C. elegans* embryos; also the epiblast of mammals.

fractal: structure with similar patterns recurring on different scales.

functional stem cells: see actual stem cells.

G_0: indefinite postmitotic gap.

G_1: specific postmitotic gap.

G_2: specific pretmitotic gap.

gametes: generally germ or sex cells of algae, animals, fungi, and plants not necessarily differentiated into eggs and spermatozoa. See germ cells.

genetic engineering (also **manipulation):** the manufacture of genes through recombinant DNA technology and their introduction into the germ line of organisms.

genetic polymorphism: the presence of two or more forms of a gene.

genome: the species-specific census of genes.

genotype: the sum of all genes in an individual; an organism's specific set of genes; characteristic set of similar or dissimilar alleles for a given gene in diploid organisms.

germ (or sex cells): cells capable of launching a new generation by fusing in an act of fertilization or conjugation; haploid cells capable of participating in fertilization; typically eggs and spermatozoa in animals but also pollen in plants or more generally gametes. Also fertilized egg, zygote, and cells formed during cleavage

germinal ridge: rudimentary gonads of vertebrate embryo.

germ line: germ cells transmitted continuously through generations.

germplasm: archaic term for self-replicating nuclear genes; regulatory substances directing germ determination.

gerontic gene: gene effecting aging and no other trait.

gonad: organs containing sex cells; ovary or testis of animals.

GS: germ stem.

haploid: eukaryotic cell containing a single dose of hereditary material.

Hayflick limit: (discovered by Leonard Hayflick) number of times cells divide in tissue culture.

hemimetabolous: See exopterygote.

HIV-1: human immunodeficiency virus 1; the retrovirus associated with AIDS.

holometabolous: See endopterygote.

homeostasis: immanent ability of organisms for self-maintenance.

homologous chromosomes: any pair of more or less identical chromosomes, each originating in a different parent; chromosomes that line up, and exchange parts prior to reduction in meiosis.

Homo sapiens: the Latinate name of the species including all human beings.

HSC: hematopoietic stem cell.

ICM: inner cell mass of blastocyst.

indeterminacy: the phenomenon of naive cells differentiating in any of several directions.

instar (caterpillars of moths and butterflies and maggots of flies): larval stages of endopterygote insect development prior to pupa.

Ig: immunoglobulin; antibody.

in vitro: (literally "in glass") generally, tissue culture or raising cells in plastic containers.

in vitro fertilization (IVF): fertilization that takes place in plastic container followed by transfer of the resulting blastocyst (known as embryo transfer) back to the uterine tube or uterus of the egg's biological parent or a surrogate parent (aka prenatal foster parent).

in vivo: (literally "in life") generally, within an organism; transferring cells among organisms.

kilobases: a length of DNA containing the bases of one thousand nucleotide pairs.

LIF: leukemia inhibitory factor; promotes growth of ES and GS cells in tissue culture.

LRC: label-retaining cell; thought to be a stem cell especially in the "bulge" region of hair follicles.

life expectancy: average duration of life after reaching given age.

life span: the duration of an individual's life from fertilization to death; the average duration of life among members of a species.

life tables: Tables of vital statistics organized by years, and consisting of columns, and rows of statistics, sometimes combined, for convenience, in groups of year.

lifetime: the interval between birth and death.

mean: the value obtain by dividing the sum of all values in a distribution by the number of values in the distribution; the average value in a distribution.

medium or **midpoint:** the value in a distribution halfway between the greatest and the least value.

methylation: addition of a methyl group (CH_3–) to a larger molecule, typically DNA, in the case of gene silencing.

meiosis: meaning "to cut in half" but referring to the reduction by half of nuclear chromosomes; typically the series of two nuclear divisions following one act of DNA replication resulting in producing haploid nuclei from diploid nuclei.

mode: the value in a distribution appearing most frequently.

mRNA: messenger RNA.

Mus musculus: the Latinate name of a species of rodents including laboratory mice, frequently considered a model system for aging research.

natural selection: the consequences of certain organisms leaving more offspring than other organisms; differential breeding resulting from differences or changes in the environment; the mechanism of Darwinian evolution.

neoteny: slowing of animal development leading to retention of juvenile features in sexually mature state.

niche: where an organism makes its living.

normal distribution (bell-shaped curve): the distribution of a continuous variable in which the mean is surrounded symmetrically by standard deviations; a distribution of a discrete variable resembling a bell-shaped curve in which the mean, median, and mode are virtually identical.

nymph: sexually immature larva of exopterygote insects.

ontogeny: the development of an organism.

oocytes: female germ-line cells between the beginning of meiosis and the completion of their second meiotic division.

oogenesis: the development and release of eggs.

organ: structures within an organism made of tissues and functioning, in both normal and disease states, at the behest of their tissues and, hence, of their cells; sometimes classified as **indigenous** (made by local tissues), **stratified** (made by the fusion of layers of tissues), or **colonized** (consisting of a local tissue matrix invaded, and taken over functionally by hematopoietic or germ line cells).

parenchyma: typically the epithelial component of an organ, but sometimes used for the dominant cell or tissue type within an organ (for example, lymphocytes in the thymus).

PCD: programmed cell death.

PGC: primordial germ cell.

phenotype: the sum of all traits in an individual; an organism's specific set of characteristics.

phylogeny: the evolution of a species (race).

plasticity: the ability of previously committed cells to differentiate along new pathways.

pleiotropy: multiple effects of a single gene.

pluripotent (pleuripotency): ability of clonally derived cell lines to differentiate into a variety of cell types.

polygenes or **multiple factors:** groups of genes with additive effects.

preembryo: see blastocyst.

***Proales decipiens*:** the Latinate name of a species of rotifers; largely freshwater, small, free-living multicellular animals with cilia around their mouth that seem to rotate and drive food into the gullet.

progeroid syndromes: premature onset of symptoms associated with aging.

prokaryote: synonym for noncompartmentalized life forms (Archaea and Bacteria).

pronucleus: one of two nuclei (female or male) within zygote.

pupa: the stage in the lifecycle of endopterygote insects between instar and adult during which metamorphosis occurs, typically within a chrysalis.

quiescent parenchyma: cells of expanding tissues capable of reentering cell cycle under stress and contributing to regeneration.

RAG-1: recombination-activating gene-1, required for the rearrangement of immunoglobulin gene segments.

recapitulation: the notion that development (ontogeny) repeats by compression and terminal addition the evolutionary stages (phylogeny) of a species (race).

regenerative adult tissue: see steady-state adult tissue.

replicative senescence: loss of ability to divide in body cells, typically demonstrated following isolation in tissue culture.

reserve cells: cells in static tissues and steady-state tissues that have left the cell cycle but may return to cycling under stress and contribute to regeneration; cognates of stem cells.

retrovirus: a virus utilizing ribonucleic acid (RNA) as its hereditary material containing a reverse transcriptase, an enzyme that replicates RNA as DNA.

RNA: ribonucleic acid.

ROS: reactive oxygen species.

***Saccharomyces cerevisiae (S. cerevisiae)*:** a species of yeast generally called budding or brewers' yeast and frequently considered a model system for aging research.

SAGE: serial analysis of gene expression.

set-aside cells: slowly cycling or non-cycling transit cells of embryos, larvae, and other growing organisms, including cells comprising imaginal disks in endopterygote (holometabolous) insects.

sex: broadly, the cycle of nuclear events in which the amount of hereditary material within a cell is doubled by fertilization and halved by reduction divisions during meiosis.

sex cells: see germ cells; gametes.

soma (somata, pl.): body of eukaryote exclusive of germ.

somatic lines: lineages of body cells.

somatoplasm: generally cytoplasm; determinants of somatic differentiation.

species: category of sexually reproducing organisms that tend to breed exclusively with each other; organisms resembling each other and distinguishable from other organisms either anatomically, behaviorally, and/or biochemically.

species specific: a characteristic identified exclusively with a species and constituting a taxonomic criterion.

species typical: a characteristic typically found in a species but not a taxonomic criterion.

spermatocytes: male germ-line cells between the beginning of meiosis and the completion of their second meiotic division.

SR: self-renewing (stem) cells; perform asymmetric cell division.

SSEA: stage specific embryonic antigens.

standard deviation: a value reflecting variation around a mean (variance) roughly equal to two-thirds the area on either side of the mean.

steady state (or regenerative) tissues: dynamic cell populations maintained by self-renewing stem cells, transit amplifying (TA) cells, and nonproliferative, differentiating or maturing cells.

stem cells: (sensu lato) self-renewing (SR) cells in steady state tissues, their cognates (see reserve and cache cells), and homologues in indefinitely growing organisms; (sensu stricto) SR cells of adult steady-state tissue; also proliferative embryonic, germ and pluripotent cells raised in tissue culture.

stroma: the connective tissue component of an organ, typically consisting of a capsule and plates (trabeculae, a mediasteinum or hilum) connected to vessels and ducts.

surrogate parent (aka prenatal foster parent): see in vitro fertilization.

survivorship distribution (curve): plot showing how a cohort dies out; graph showing number of organisms remaining in a cohort (on the Y axis) as a function of time (on the X axis) until the last organism is dead.

TA: transit amplifying: cell dividing symmetrically, producing clonal lineages of cells that differentiate in steady-state tissues.

telomere: the knobs at the ends of chromosomes that reduce chromosomal stickiness; the ends of linear, nuclear DNA composed of repeated sequences; ends of chromosomes eroded during replication; buffers against loss of chromosomal genes; allegedly a "count-down" timer determining the limit of cell replication (see Hayflick limit).

telomerase: an enzyme containing an RNA template and a reverse transcriptase functioning in lengthening telomeres.

tissues: composites of extracellular material and cells roughly similar in structure and function; classically **epithelial, connective, muscle, and nerve,** to which, now, **blood** and **germ cells** have been added.

transcriptome: products of all transcripts (mRNA) in cells.

transdifferentiate: the loss of cellular differentiation coupled to a change in differentiation without the intervention of cell division.

transfection: passive introduction of foreign genes into cells.

transgenic organisms: organisms with foreign genes.

trisomy: a congenital anomaly in which a chromosome (or a part of it) appears three times rather than the usual two; **trisomy 21:** an extra chromosome 21.

trophectoderm: outer layer of blastocyst cells.

zygote: fertilized egg with two haploid nuclei preparing for cleavage.

Bibliography

Abbott, A. An end to adolescence. *Nature* 433:27; Jan. 6, 2005.

Abkowitz J. L., S. N. Catlin, M. T. McCallie, and P. Guttorp. Evidence that the number of hematopoietic stem cells per animal is conserved in mammals. *Blood* 100:4679–80; 2002.

Ackermann, M., S. C. Stearns, and U. Jenal, Senescence in a bacterium with asymmetric division. *Science* 300:1920; 2003.

Aguilaniu, H., L. Gustafsson, M. Rigoulet, and T. Nyström. Asymmetric inheritance of oxidatively damaged proteins during cytokinesis. *Science* 299:1751–53; 2003.

Aitken, R. J., P. Koopman, and S. E. M. Lewis. Seeds of concern. *Nature* 432:48–52; 2004 (doi:10.1038/432048a).

Allard, M., V. Lèbre, and J-M. Robine, *Jeanne Calment: From Van Gogh's Time to Ours 122 Extraordinary Years.* Translated by Beth Coupland. New York: W.H. Freeman, 1998. (Originally published as M. Allard, *120 ans de Jeanne Calment, doyenne de l'humanité.* 1994.)

Allegrucci, C., A. Thurston, E. Lucas, and L. Young. Epigenetics and the germline. *Reproduction* 129:137–49; 2005.

Allsopp, R. C., and I. L. Weissman, Replicative senescence of hematopoietic stem cells during serial transplantation: Does telomere shortening play a role? *Oncogene* 21:3270–73; 2002.

Alonso, L., and E. Fuchs, Stem cells of the skin epithelium. *Proc. Natl. Acad. Sci. USA* 100:11830–35; 2003.

Anderson, D. J, and V. Apanius. Actuarial and reproductive senescence in a long-lived seabird: Preliminary evidence. *Exp. Gerontol.* 38:757–60; 2003.

Andrew, D., and R. Aspinall. Age-associated thymic atrophy is linked to a decline in IL-7 production. *Exp. Gerontol.* 37:455–63; 2002 (doi:10.1016/S0531–5565(01)00213–3).

Anon. US scraps study on cancer fallout from nuclear testing. *Nature* 434:688; 2005.

———. 152-year-old lake sturgeon caught in Ontario. *Commercial Fisheries Review* 16, no. 9:28; 1954.

AP. Americans are living longer. *New York Times* (March 1, 2005): A15, 3.

Arantes-Oliveira, N., J. R. Berman, and C. Kenyon. Healthy animals with extreme longevity. *Science* 302:611; 2003.

Arias, E. United States Life Tables, 2000. *National Vital Statistics Reports*, vol. 5; no. 3. Hyattsville, Md.: National Center for Health Statistics, 2002.

Ariès, P. *The Hour of Our Death.* Translated by Helen Weaver. New York: Knopf, 1981.

Arrasate, M., S. Mitra, E. S. Schweitzer, M. R. Segal, and S. Finkbeiner. Inclusion body formation reduces levels of mutant huntingtin and the risk of neuronal death. *Nature* 431:805–10; 2004.

Aspinall, R., and D. Andrew. Thymic atrophy in the mouse is a soluble problem of the thymic environment, *Vaccine*, 18:1629–37; 2000 (doi:10.1016/S0264–410X(99) 00498–3).

Astrow, S., B. Holton, and D. Weisblat, Centrifugation redistributed factors determining cleavage patterns in leech embryos. *Develop. Biol.* 120:270–83; 1987.

Auger, J., J. M. Kunstmann, F. Czyglik, P. Jouannet. Decline in semen quality among fertile men in Paris during the past 20 years. *N. Engl. J. Med.* 332:281–85; 1995.

Austad, S. N. *Why We Age: What Science Is Discovering about the Body's Journey through Life.* New York: Wiley, 1997.

Baker, T. G. "Oogenesis and Ovulation." In *Reproduction in Mammals*, book 1, *Germ Cells and Fertilization*, 2nd ed., edited by C. R. Austin and R. V. Short, 17–45. Cambridge: Cambridge University Press, 1982.

Balkwill, F., and L. M. Coussens. An inflammatory link. *Nature* 431:405–6; Sept. 23, 2004.

Bar-Dayana, Y., A. Afekb, Y. Bar-Dayanc, I. Goldberg, and J. Kopolovic. Proliferation, apoptosis and thymic involution. *Tissue and Cell* 31:391–96; 1999 (doi:10.1054/tice.1999.0001).

Bell, G. *Sex and Death in Protozoa: The History of an Obsession.* Cambridge: Cambridge University Press, 1988.

Bell, G., and V. Koufopanou. "The Cost of Reproduction." In *Oxford Surveys in Evolutionary Biology*, vol. 3, edited by R. Dawkins and M. Ridley, 83–131. Oxford: Oxford University Press, 1986.

Beltrami, A. P., L. Barlucchi, D. Torella, M. Baker, F. Limana, S. Chimenti, H. Kasahara, et al. Adult cardiac stem cells are multipotent and support myocardial regeneration. *Cell* 114:763–76; 2003.

Benecke, M. *The Dream of Eternal Life: Biomedicine, Aging, and Immortality.* Translated by Rachel Rubenstein. New York: Columbia University Press, 2002.

Bergson, H. *Creative Evolution.* Translated by Arthur Mitchell. Mineola: Dover Publications, 1998. (Originally published in 1911.)

Berlekamp, E. R., J. H. Conway, R. K. Guy. *Winning Ways: For Your Mathematical Plays.* Vol. 2, *Games in Particular.* London: Academic Press, 1982.

Berrill, N. J. *Growth, Development and Pattern.* San Francisco: W. H. Freeman, 1961.

Bertsch, S., and F. Marks. Effects of foetal calf serum and epidermal growth factor on DNA synthesis in explants of chick embryo epidermis. *Nature* 251:517–19; 1974.

Blackford, R. "Should We Fear Death? Epicurean and Modern Arguments," In *The Scientific Conquest of Death: Essays on Infinite Lifespans*, edited by Immortality Institute, 257–69. Buenos Aires: LibrosEn Red, 2004.

Blackler, A. W. Transfer of primordial germ cells between two subspecies of *Xenopus laevis. J. Embryol. exp. Morphol.* 10:641–51; 1962.

Blackler, A. W., and M. Fischbert. Transfer of primordial germ cells in *X. laevis. J. Embryol. exp. Morphol.* 9:634–41; 1961.

Blackler, A. W., and C. A. Gecking. Transmission of sex cells of one species through the body of a second species in the genus *Xenopus.* I. Intraspecific matings. II. Interspecific matings. *Develop. Biol.* 27:376–84, 385–94; 1972.

Blau, H. M., T. R. Brazelton, and J. M. Weimann. The evolving concept of a stem cell: Entity or function? *Cell* 105:829–41; 2001.

Bodey, B., B. Bodey, Jr., S. E. Siegel, and H. E. Kaiser. Involution of the mammalian thymus, one of the leading regulators of aging. *In Vivo* 11:421–40; 1999.

Bodnar, A.G., M. Ouellette, M. Frolkis, S. E. Holt, C-P. Chiu, B. B. Morin, C. B. Harley, J. W. Shay, S. Lichtsteiner, and W. E. Wright. Extension of life-span by introduction of telomerase into normal human cells. *Science* 279:349–52; 1998.

Boggs, D. R., D. F. Saxe, and S. S. Boggs. Aging and hematopoiesis. II. The ability of bone marrow cells from young and aged mice to cure and maintain cure in W/Wv. *Transplantation* 37:300–6; 1984.

Boklage. C. E. "The Biology of Human Twinning: A Needed Change of Perspective." In *Multiple Pregnancy: Epidemiology, Gestation and Perinatal Outcome*, edited by I. Blickstein, L. G. Keith, and E. Papiernik, 41–50. New York: Parthenon, 2005.

Boldsen, J. L., and R. R. Paine. "The Evolution of Human Longevity from the Mesolithic to the Middle Ages: An Analysis Based on Skeletal Data." In *Exceptional Longevity: From Prehistory to the Present*, edited by B. Jeune and J. W. Vaupel, 25–36. Monographs on Population Aging, 2. Odense: Odense University Press, 1995.

Bonhoeffer, S., C. Chappey, N.T. Parkin, J. M. Whitcomb, and C. J. Petropoulos. Evidence for positive epistasis in HIV-1. *Science* 306:1547–51; 2004.

Booke of Psalme, The. *The King James Bible*, 1996–2006 ProQuest Information and Learning Company. http://collections.chadwyck.com/

Boulter, M. *Extinction: Evolution and the End of Man*. New York: Columbia University Press, 2002.

Bowler, P. J. *Life's Splendid Drama: Evolutionary Biology and the Reconstruction of Life's Ancestry 1860–1940*. Chicago: University of Chicago Press, 1996.

Bowler, P. J. *The Non-Darwinian Revolution: Reinterpreting a Historical Myth*. Baltimore: Johns Hopkins University Press, 1988.

Brasier, M., and J. Antcliffe. Decoding the Ediacaran enigma. *Science* 305:1115–17; 2004.

Brawley, C., and E. Matunis. Regeneration of male germline stem cells by spermatogonial dedifferentiation in vivo. *Science* 304:1331–34; 2004.

Brelinska, R. Thymic epithelial cells in age-dependent involution. *Microsc. Res. Tech.* 62:488–500; 2003.

Bromage, T. G., and M. C. Dean. Re-evalutaion of the age at death of immature fossil hominids, *Nature* 317:525–27; 1985.

Bukovsky, A., M. Svetlikova, and M. R. Caudle. Oogenesis in Cultures Derived from Adult Human Ovaries. *Repro. Biol. Endocrinol.* 2005, 3:17 (doi:10.1186/1477–7827–3–17). http://www.rbej.com/content/3/1/17.

Butler, D. The fertility riddle. *Nature* 432:38–39; 2004 (doi:10.1038/432038a).

———. US abandons health study on agent orange. *Nature* 434:687; 2005.

Cadbury, D. *The Feminization of Nature: Our Future at Risk.* London: Penguin, 1997.

Cailliet, G. M., A. H. Andrews, E. J. Burton, D. L. Watters, D. E. Kline, and L. A. Ferry-Graham. Age determination and validation studies of marine fishes: Do deep-dwellers live longer? *Exp. Gerontol.* 6:739–64; 2001.

Cairns, J., J. Overbaugh, and S. Miller. The origin of mutants. *Nature* 335:142–45; 1988.

Caldwell, J. C. Toward a restatement of demographic transition theory. *Population Development and Review* 2:321–66; 1976.

Camazine, S., L.-L. Deneubourg, N. R. Franks, J. Sneyd, G. Theraulaz, and E. Bonabeau. *Self-Organization in Biological Systems.* Princeton: Princeton University Press, 2001.

Carey, J. R. *Longevity: The Biology and Demography of Life Span.* Princeton: Princeton University Press, 2003.

Carlsen E., A. Giwercman, N. Keiding, and N. E. Skakkebaek. Evidence for decreasing quality of semen during past 50 years. *Brit. Med. J.* 305:609–13; 1992.

Castro, R. F., K. A. Jackson, M. A. Goodell, C. S. Robertson, H. Liu, H. D. Shine. Failure of bone marrow cells to transdifferentiate into neural cells in vivo. *Science* 297:1299; 2002.

Centers for Disease Control and Prevention (CDC), "Linked Birth/Infant Death Data Set." http://www.cdc.gov/nchs/data/hus/tables/2003/03hus(add table number 001 to 151).pdf/

———. "Interactive Atlas of reproductive Health: Home." Department of Health and Human Services, http://www.cec.gov/reproductivehealth/gisatlas (2004).

Chakravarti, B., and G. N. Abraham. Review: Aging and T-cell-mediated immunity *Mech. Ageing Dev.* 108:183–206; 1999 (doi:10.1016/S0047–6374(99)00009–3).

Charnov, E. L. Evolution of life history variation among female mammals. *Proc. Natl. Acad. Sci. USA* 88:1134–37; 1991.

Chen J., C. M. Astle, and D. E. Harrison. Hematopoietic senescence is postponed and hematopoietic stem cell function is enhanced by dietary restriction. *Exp. Hematol.* 31:1097–103; 2003.

Chiang, C. L., *The Life Table and Its Applications.* Malabar, Fla.: Robert E. Krieger, 1984.

Christie, A. *Curtain: Hercule Poirot's Last and Greatest Case.* New York: Kangaroo Book, 1975.

———. *The Man in the Brown Suit,* 3rd ed. Hammersmith: HarperCollins, 1997.

Clark A. T., M. S. Bodnar, M. Fox, R. T. Rodriquez, M. J. Abeyta, M .T. Firpo, and R. A. Pera. Spontaneous differentiation of germ cells from human embryonic stem cells in vitro. *Hum. Mol. Genet.* 13:727–39; 2004.

Clark, J. M., and E. M. Eddy. Fine structural observations on the origin and associations of primordial germ cells of the mouse. *Develop. Biol.* 47:136–55; 1975.

Clarke, A. C. *Greetings, Carbon-based Bipeds! Collected Essays, 1934–1998.* Edited by Ian T. Macauley. New York: St. Martin's Press, 1999.

Clermont, Y. Quantitative analysis of spermatogenesis in the rat. A revised model for renewal of spermatogonia. *Am. J. Anat.* 111:111–29; 1962.

Coffer, P. OutFOXing the grim reaper: Novel mechanisms regulating longevity by Forkhead transcription factors. *Sci. STKE* 2003, pe39; 2003.

Cohen, H. Y., C. M. Miller, K. J. Bitterman, N. R. Wall, B. Hekking, B. Kessler, K. T. Howitz, M. Corospe, R. deCabo, and D. A. Sinclair. Calorie restriction promotes mammalian cell survival by inducing the SIRT1 deacetylase. *Science* 305:390–92; 2004.

Coles, L. S. Co-founder Los Angeles Gerontology Research Group. Personal communication, May 18, 2004.

———. Short Table of Cumulative Supercentenarians by Age: e-mail to Gerontology Research Group mailing list, March 12, 2005, http://lists.ucla.edu/cgi-bin/mailman/listinfo/grg.

Comfort, A. *The Biology of Senescence.* Edinburgh: Churchill Livingstone, 1979.

Comins, N. F. *What If the Moon Didn't Exist?* New York: Harper Perennial Library; Reprint; 1993.

Centers for Disease Control and Prevention (CDC): http://www.cdc.gov/reproductive-health/gisatlas (2004).

Congdon, J. D., R. D. Nagle, O. M. Kinney, R. C. van Loben Sels, T. Quinter, and D. W. Tinkle. Testing hypotheses of aging in long-lived painted turtles (*Chrysemys picta*). *Exp. Gerontol.* 38:765–72; 2003.

Conklin, E. G. "Cleavage and differentiation." In *Defining Biology: Lectures from the 1890s,* edited by J. Mainenschein, 151–77. Cambridge: Harvard University Press, 1986.

Conway Morris, S. *Life's Solution: Inevitable Humans in a Lonely Universe.* Cambridge: Cambridge University Press, 2003.

Coqueugniot, N., J-J. Hublin, F. Vellion, F. Housët, and T. Jacob. Early brain growth in *Homo erectus* and implications for cognitive ability. *Nature* 431:299–302; 2004.

Costello, D. P., and C. Henley. Spiralian development: A perspective. *Am. Zool.* 16:277–91; 1976.

Courier Mail. "Older dads not in the swim." http://www.thecouriermail.news.com.au/common/story_page/0,5936,9386234%5E953,00.html; April 26, 2004.

Courtillot, V. *Evolutionary Catastrophes: The Science of Mass Extinction.* Translated by Joe McClinton. Cambridge: Cambridge University Press, 1999.

Couzin, J. Reproductive biology. Textbook rewrite? Adult mammals may produce eggs after all. *Science* 303:1593; 2004

Cowan, C. A., I. Klimanskaya, J. McMahon, J. Atienza, J. Witmyer, J. P. Zucker, S. Wang, et al. Derivation of embryonic stem-cell lines from human blastocysts. *N. Engl. J. Med.* 350:13 at www.nejm.org. March 25, 2004.

Crawford, R. M. "Some Considerations of Size Reduction in Diatom Cell Walls." In *Proceedings of the Sixth Symposium on Recent and Fossil Diatoms*, edited by R. Ross, 253–65. Koenigstein: Koeltz, 1981.

Cuervo, A. M., L. Stefanis, R. Fredenburg, P. T. Lansbury, D. Sulzer. Impaired degradation of mutant a-synuclein by chaperone-mediated autophagy. *Science* 305:1292–95; 2004.

Cutler, R. G. "Life-Span Extension." In *Aging: Biology and Behavior*, edited by James L. McGaugh and Sara B. Kiesler, 31–76. New York: Academic Press, 1981.

Dalton, R. Fresh study questions oldest traces of life in Akilia rock. *Science* 429:688; June 17, 2004.

Damjanov, I., A. Damjanov, and D. Solter. "Production of Teratocarcinomas from Embryos Transplanted to Extra-Uterine Sites." In *Teratocarcinomas and Embryonic Stem Cells: A Practical Approach*, edited by E. J. Robertson, 1–18. Oxford: IRL Press, 1987.

Darwin, F. *The Life of Charles Darwin.* London: Senate, 1995. (Originally published in 1902.)

Davenport, R. J. Fanning the flames: Telomerase aggravates oxidative damage in mitochondria *Sci. Aging Knowl. Environ.* 2004a Oct. 6; 2004(40):nf90.

———. Paying the price. *Sci. Aging Knowl. Environ.* 2004b Dec. 1; 2004(48):nf107.

Davidson, E. H. Lineage-specific gene expression and the regulative capacities of the sea urchin embryo. *Development* 105:421–45; 1989.

Davis, G. E. Jr., and W. E. Lowell. The Sun determines human longevity: Teratogenic effects of chaotic solar radiation. *Med. Hypotheses* 63:574–81 (abstract); 2004.

Dean, M. C. Of faster brains and bigger teeth. *Nature* 330:213; 1987.

de Beer, G. R. *Embryos and Ancestors.* Oxford: Clarendon Press, 1951.

de Grey, A. Dr. Cynthia Kenyon in March Discover Magazine: e-mail to Gerontology Research Group mailing list, Feb. 14, 2004a, http://lists.ucla.edu/cgi-bin/mailman/listinfo/grghttp://lists.ucla.edu/cgi-bin/mailman/listinfo/grg.

————. New news article about the human life span—from New Zealand: e-mail to Gerontology Research Group mailing list, May 3, 2004b, http://lists.ucla.edu/cgi-bin/mailman/listinfo/grghttp://lists.ucla.edu/cgi-bin/mailman/listinfo/grg.

————. "The War on Aging" In *The Scientific Conquest of Death: Essays on Infinite Lifespans*, edited by Immortality Institute, 29–45. Buenos Aires: LibrosEn Red, 2004c.

De Haan, G., and G. van Zant. Genetic analysis of hemopoietic cell cycling in mice suggests its involvement in organismal life span. *FASEB J.* 13:707–13; 1999.

de Lange, T. Telomeres and senescence: Ending the debate. *Science* 279:334–35; 1998.

DeLillo, D. *Americana.* London: Penguin, 1990.

de Magalhães, J. P. "The dream of Elixir Vitae" In *The Scientific Conquest of Death: Essays on Infinite Lifespans*, edited by Immortality Institute, 47–62. Buenos Aires: LibrosEn Red, 2004.

Demeny, P. Early fertility decline in Austria-Hungary: A lesson in demographic transition. *Daedalus* 97:502–22; 1968.

Dennis Smith, L. A test of the capacity of presumptive somatic cells to transform into primordial germ cells in the Mexican axolotl. *J. Exp. Zool.* 156:229–42; 1964.

DePaolo, L. V. "Dietary Modulation of Reproductive Function." In *Modulation of Aging Processes by Dietary Restriction*, edited by Byung Pal Yu, 221–45. Boca Raton: CRC Press, 1994.

Derrida, J. *The Gift of Death.* Translated by David Wills. Chicago: University of Chicago Press, 1992.

Diamond, J. Causes of death before birth. *Nature* 329:487–88; October 1987.

Doetsch, F. The glial identity of neural stem cells. *Nature Neurosci.* 6:1127–34; 2003.

Dominguez-Gerpe, L., and M. Rey-Mendez. Age-related changes in primary and secondary immune organs of the mouse. *Immunol. Invest.* 27:153–65; 1998.

Donnell, D. M., L. S. Corley, G. Chen, and M. R. Strand. Caste determination in a polyembryonic wasp involves inheritance of germ cells. *Proc. Natl. Acad. Sci. USA* 101:10095–100; 2004.

Doolittle, W. F. Phylogenetic classification and the Universal Tree. *Science* 284:2124–28; 1999.

Dooren, J. C. Americans' life expectancy rose to record high in 2003. *Wall Street Journal*, (March 1, 2005): A1, D8.

Dor, Y., J. Brown, O. I. Martinez, and D. A. Melton. Adult pancreatic β-cells are formed by self-duplication rather than stem-cell differentiation. *Nature* 429:41–46; 2004.

Douglas, Lady John Montague, *Annie Lawrie.* Buffalo: J Sage & Sons; c1855.

Driver, C. Where is the somatic mutation that causes aging? *BioEssay* 26:1160–63; 2004.

Drug Enforcement Administration Museum. "DEA History: 1985–1990: The Crack Epidemic" See www.streetdrugs.org/ follow links to crack cocaine. (2004).

Duque-Parra J. E. Neurobiological relations and aging. *Rev. Neurol.* 36:549–54 (abstract in English); 2003.

Eaton, S. M., E. M. Burns, K. Kusser, T. D. Randall, and L. Haynes. Age-related defects in CD4 T cell cognate helper function lead to reductions in humoral responses. *J. Exp. Med.* 200:1613–22; 2004.

Eddy, E. M. Germ plasm and the differentiation of the germ cell line. *Int. Rev. Cytol.* 43:229–80; 1975.

Eddy, E. M., J. M. Clark, D. Gong, and B. A. Fendersen. Origin and migration of primordial germ cells in mammals. *Gamete Res.* 4:333–62; 1981.

Eddy. E. M., and A. C. Hahnel. "Establishment of the Germ Cell Line in Mammals," In *Current Problems in Germ Cell Differentiation*, 7th Symposium of the British Society for Developmental Biology, edited by A. McLaren and C. C. Wylie, 41–69. Cambridge: Cambridge University Press, 1983.

Eiler, J. M., S. J. Mojzsis, G. Arrhenius. Carbon isotope evidence for early life. *Nature* 386:665; 1997.

eNature. "Survivor!" National Wildlife Federation, http://www.enature.com/articles/ (Nov. 12, 2003).

Endler, J. A. *Natural Selection in the Wild.* Princeton: Princeton University Press, 1986.

Epistle of Pavl, The: *The Apostle to the Romanes.* 1996–2003 ProQuest Information and Learning Company. http://collections.chadwyck.com/.

Erickson, G. M., P. J. Makovicky, P. J. Currie, M. A. Norell, S. A. Yerby, and C. R. Brochu. Giantism and comparative life-history parameters of tyrannosaurid dinosaurs. *Nature* 430:772–77; 2004.

Espejel, S., M. Martín, P. Klatt, J. Martín-Caballero, J. M. Flores, and M. A. Blanco. Shorter telomeres, accelerated ageing and increased lymphoma in DNA-PGCs-deficient mice. *EMBO reports* AOP; 2004 (doi:10.1038/sj.embor.7400127 23 April).

Evans, M. J., and M. H. Kaufman. Establishment in culture of pluripotential cells from mouse embryos. *Nature* 292:154–56; 1981.

Extavour, C. G. Hold the germ cells, I'm on duty. *BioEssays* 26:1263–67; 2004.

Extavour, C. G., and M. Akam. Mechanisms of germ cell specification across the metazoans: Epigenesis and preformation. *Development* 130:5869–84; 2003; doi:10.1242/dev.00804.

Fabrizio, P., L. Battistella, R. Vardavas, C. Gattazzo, L.-L. Liou, A. Diaspro, J. W. Dossen, E. B. Gralla, and V. D. Longo, Superoxide is a mediator of an altruistic aging program in *Saccharomyces cerevisiae*, *J. Cell Biol.* 166:1055–67; September 27, 2004 (doi: 10.1083/jcb.200404002).

Fedo, C. M., and M. J. Whitehouse. Metaxomatic origin of quartz-pyroxene rock, Akilia, Greenland, and implications for Earth's earliest life. *Science* 296:1448–52; 2002.

Fehr, H., S. Jokisch, and L. J. Kotlikoff. Mortality, and the developed world's demographic transition. *CESifo* Working Paper No. 1326:1–30; 2004.

Finch, C. E. The age-old problem of mortality. *Nature* 428:125; 2004.

————. *Longevity, Senescence, and the Genome.* Chicago: University of Chicago Press, 1990.

Finch, C. E., and T. B. L. Kirkwood. *Chance, Development and Aging.* New York: Oxford University Press, 2000.

Finch, C. E., and E. M. Crimmins. Inflammatory exposure and historical changes in human life-spans. *Science* 305:1736–39; 2004.

Flew, A. Introduction. In *An Essay on the Principle of Population and A Summary View of the Principle of Population,* by T. R. Malthus, edited by Antony Flew, 7–56. London: Penguin, 1970. (Originally published in 1798.)

Forbes, S. N., R. K. Valenzuela, P. Keim, and P. M. Service. Quantitative trait loci affecting life span in replicated populations of *Drosophila melanogaster.* I. Composite interval mapping. *Genetics* 168:301–11; 2004 (doi: 10.1534/genetics.103.023218).

Foucault, M. *The History of Sexuality.* Vol. 1, *An Introduction.* Translated by Robert Hurley. New York: Vintage Books, 1980.(Originally published as *La Volenté de savoir.* Paris: Gallimard, 1976.)

Franceschi, C., D. Monti, D. Barbieri, S. Salvioli, E. Grassilli, M. Capri, L. Troiano, et al. Successful immunosenescence and the remodeling of immune responses with ageing. *Nephrol. Dial. Transplant.* 11 Suppl. 9:18–25; 1996.

Franceschi, C. M. Bonafè, and S. Valensin. Human immunosenescence: The prevailing of innate immunity, the failing of clonotypic immunity, and the filling of immunological space. *Vaccine* 18:1717–20; 2002 (doi:10.1016/S0264-410X(99)00513-7).

Frank, P. W. Life span: Human life span: Actual versus possible life span. *Encyclopedia Britannica,* from *Encyclopedia Britannica,* 2005 DVD. Copyright © 1994–2004 Encyclopedia Britannica, Inc. May 30, 2004.

Freitas, R. A., Jr. "Nanomedicine." In *The Scientific Conquest of Death: Essays on Infinite Lifespans,* edited by Immortality Institute, 77–92. Buenos Aires: LibrosEn Red, 2004.

Fukuyama, F. *The End of History and the Last Man.* New York: Free Press, 1992.

Fukuyama, F. *Our Posthuman Future: Consequences of the Biotechnology Revolution.* New York: Farrar, Straus and Giroux, 2002.

Furnes, II., N. R. Banerjee, K. Muehlenbachs, H. Staudigel, and M. de Wit. Early life recorded in Archean pillow lavas. *Science* 304:578–81; 2004.

Gage, F. H. Mammalian neural stem cells. *Science* 287:1433–38; 2000.

Gardner, R. L. "Developmental Potency of Normal and Neoplastic Cells of the Early Mouse Embryo." In *Birth Defects,* edited by J. W. Littlefield and J. Grouchy. Excerpta Medica International Congress Series No. 432, 154–166. Amsterdam: Excerpta Medica, 1978.

Garstang, W. *Larval Forms: And Other Zoological Verses.* With an introduction by Sir Alister Hardy, and a foreword by Michael LaBarbera. Chicago: University of Chicago Press, 1985.

Gavrilov, L. A., and N. S. Gavrilova, *The Biology of Life Span: A Quantitative Approach.* Revised and updated English Edition. Edited by V. P. Skulachev and translated by John Payne and Liliya Payne. Chur: Harwood Academic Publishers; 1991. (Originally published in Russian, Moscow: Nauka, 1986).

————. Evolutionary Theories of Aging and Longevity. *TheScientificWorldJOURNAL* 2:339–56; 2002 (doi 10.1100/tsw.2002.96).

————. Re: Evolutionary theories of aging and mortality deceleration: e-mail to Gerontology Research Group mailing list, April 16, 2005, http://lists.ucla.edu/cgi-bin/mailman/listinfo/grg.

Geddes, M. "An Introduction to Immortalist Morality." In *The Scientific Conquest of Death: Essays on Infinite Lifespans*, edited by Immortality Institute, 239–55. Buenos Aires: LibrosEn Red, 2004.

Geiger, H., S. Sick, C. Bonifer, A. M. Müller. Globin gene expression is reprogrammed in chimeras generated by injecting adult hematopoietic stem cells into mouse blastocysts, *Cell* 93:1055–65; 1998.

Geijsen, N., M. Horoschak, K. Kim, J. Gribnau, K. Eggan, and G. Q. Daley. Derivation of embryonic germ cells and male gametes from embryonic stem cells. *Nature* 427:148–54; 2004.

Genes/Interventions (Database). http://sageke.sciencemag.org/cgi/genesdb; 2004

Genesis. *The King James Bible*, 1996–2003 ProQuest Information and Learning Company. http://collections.chadwyck.com/.

George, J. C., J. Bada, J. Zeh, L. Scott, S. E. Brown, T. O'Hara, and R. Suydam. Age and growth estimates of bowhead whales (*Balaena mysticetus)* via aspartic acid racemization. *Can. J. Zool.* 77:571–80; 1999.

Gilbert, W. The exon theory of genes. *Cold Spring Harbor Symposia on Quantitative Biology* 52:901–5; 1987.

————. Genes-in-pieces revisited. *Science* 228:823–24; 1985.

———— "A Vision of the Grail." In *The Code of Codes: Scientific and Social Issues in the Human Genome Project*, edited by D. J. Kevles and L. Hood, 83–97. Cambridge: Harvard University Press, 1992.

————. The RNA world. *Nature* 319:618; 1986.

Gilbert, W., M. Long, C. Rosenberg, and M. Glynias. "Tests of the Exon Theory of Genes." In *Tracing Biological Evolution in Protein and Gene Structure*, edited by M. Gø and P. Schimmel, 237–47. Amsterdam: Elsevier, 1995.

Gilbert, W., S. J. De Souza, and M. Long. Origin of Genes. *Proc. Natl. Acad. Sci. USA* 94:7698–7703; 1997.

Gins, M., and Arakawa. *Architectural Body.* Tuscaloosa: University of Alabama Press, 2002.

Gleick, J. *Chaos: Making a New Science.* New York: Penguin, 1987.

Global Population Profile: 2000. U.S. Census Bureau; 2002. http://www.census.gov/ipc/www/wp02.html

Globerson, A. Hematopoietic stem cells and aging. *Exp. Gerontol.* 34:134–46; 1999.

Godfrey L. R., and M. R. Sutherland. Paradox of peramorphic paedomorphosis: Heterochrony and human evolution. *Am. J. Phys. Anthropol.* 99:17–42; 1996.

Gould, S. J. Of coiled oysters and big brains: How to rescue the terminology of heterochrony, now gone astray. *Evol. Dev.* 2:241–48; 2000.

———. *Ontogeny and Phylogeny.* Cambridge: Belknap Press, 1977.

Grounds, M. D., J. D. White, N. Rosenthal, and M. A. Bogoyevitch. The role of stem cells in skeletal and cardiac muscle repair. *J. Histochem. Cytochem.* 50:589–610; 2002.

Guarente, L. *Ageless Quest: One Scientist's Search for Genes that Prolong Youth.* Cold Spring Harbor: Cold Spring Harbor Laboratory Press, 2003.

Hageman, S. J. Complexity generated by iteration of hierarchical modules in Bryozoa. *Integr. Comp. Biol.* 43:87–98; 2003.

Hall, S. S. *Merchants of Immortality: Chasing the Dream of Human Life Extension.* Boston: Houghton Mifflin, 2003.

Hanson, E. D. *The Origin and Early Evolution of Animals.* Middletown: Wesleyan University Press, 1977.

Hanyu, N., Y. Kuchino, S. Nishimur, and H. Beier. Dramatic events in ciliate evolution: Alteration of UAA and UAG termination codons to glutamine codons due to anticodon mutations in *Tetrahymena* tRNA^{Gln}. *EMBO J.* 5:1307–11; 1986.

Hardt, M., and A. Negri. *Empire.* Cambridge: Harvard University Press, 2000.

Harris, R. G., E. L. Herzog, E. M. Bruscia, J. E. Grove, J. S. Van Arnam, D. S. Krause. Lack of a fusion requirement for development of bone marrow-derived epithelia. *Science* 305:90–93; 2004.

Hasty, P., J. Campisi, J. Hoeijmakers, H. van Steeg, and J. Vijg. Aging and Genome Maintenance: Lessons from the Mouse? *Science* 299:1355–59; 2003.

Hawkes, K. Human longevity: The grandmother effect. *Nature* 428:128–29; 2004 (doi:10.1038/428128a).

Hayflick, L. *How and Why We Age.* New York: Ballantine Books, 1996.

Hayflick, L. Modulating aging, longevity determination, and the diseases of old age. In *Modulating Aging and Longevity*, edited by S. I. S. Rattan, 1–15. Dordrecht, Netherlands: Kluwer Academic Publishers, 2003.

Heffner, L. J. Advanced maternal age—how old is too old? *N. Engl. J. Med.* 351:1927–29, 2004.

Heidegger, M. "The Question Concerning Technology." In *The Question Concerning Technology and Other Essays*, translated and with an introduction by William Lovitt, 3–35. New York: Harper Torchbooks, 1977.

Hekimi, S., and L. Guarente. Genetics and the specificity of the aging process. *Science* 299:1351–54; 2003.

Henry, P. F. The eastern box turtle at the Patuxent Wildlife Research Center 1940s to the present: Another view. *Exp. Gerontol.* 38:773–76; 2003.

Herskind, A. M., M. McGue, N. V. Holm, T. I. A. Sørensen, B. Harvald, and J. W. Vaupel. The heritability of human longevity: A population-based study of 2872 Danish twin pairs born 1870–1900. *Hum. Genet.* 97:319–23; 1996.

Hind, R. A. *Ethology: Its Nature and Relations with Other Sciences.* New York: Oxford University Press, 1982.

Ho, M-W. *The Rainbow and the Worm: The Physics of Organisms.* 2nd ed. Singapore: World Scientific Publishing, 1998.

Holden, C. Tiniest vertebrate (Random Samples). *Science* 305:472; July 23, 2004.

Holmes, D. J, S. L. Thomson, J. Wu, and M. A. Ottinger. Reproductive aging in female birds. *Exp. Gerontol.* 38:751–56; 2003.

Holtzer, H. Comments following L. G. Lajtha. Stem cell concepts. *Differentiation* 14:23–43; 1979.

Höstadius, S. *The Neural Crest.* Oxford: Oxford University Press, 1950.

Houthoofd, K., B. P. Braeckman, T. E. Johnson, J. R. Vanfleteren. Extending life-span in *C. elegans. Science* 1238–39; 2004.

Howard, A., and S. R. Pelc. Nuclear incorporation of P^{32} as demonstrated by autoradiographs. *Exp. Cell Res.* 2:178–87; 1951.

Howell, J. C., W-H. Lee, P. Morrison, J. Zhong, M. C. Yoder, and E. F. Srour. Pluripotent stem cells identified in multiple murine tissues. *Annals New York Academy Sciences* 996:158–73; 2003.

Hoyer, P. B. Can the Clock Be Turned Back on Ovarian Aging? *Sci. Aging Knowl. Environ.* pe11 2004; (doi: 10.1126/sageke.2004.10.pe11).

Hubbard, R. "The Theory and Practice of Genetic Reductionism—From Mendel's Laws to Genetic Engineering." In *Towards a Liberatory Biology*, edited by Steven Rose, 62–78. London: Allison and Busby, 1982.

Hübner, K, G. Fuhrmann, L. K. Christenson, J. Kehler, R. Reinbold, R. De Las Fuente, J. Wood, J. F. Strauss III, M. Boiani, and H. R. Schöler. Derivation of oocytes from mouse embryonic stem cells. *Science* 300:1251–56; 2003.

Human Mortality Database. http://www.mortality.org/

Hyman, L. H. *The Invertebrates: Protozoa through Ctenophora.* New York: McGraw-Hill, 1940.

———. *The Invertebrates*, vol. 5, *Smaller Coelomate Groups.* New York: McGraw-Hill, 1959.

Irvine, S., E. Cawood, D. Richardson, E. MacDonald, and J. Aitken. Evidence of deteriorating semen quality in the United Kingdom: Birth cohort study in 577 men in Scotland over 11 years. *Brit. Med. J.* 312:467–71; 1996.

Jackson, K. A., S. M. Majka, H. Wang, J. Pocius, C. J. Hartley, M. W. Majesky, M. L. Entman, L. H. Michae, K. K. Hirschi, and M. A. Goodell. Regeneration of ischemic cardiac muscle and vascular endothelium by adult stem cells. *J. Clin. Invest.* 107, no. 11:1395–402; 2001.

Jacob, F. *The Logic of Life: A History of Heredity.* Translated by Betty E. Spillmann. Princeton: Princeton University Press, 1973.

Jacobson, M. Clonal organization of the central nervous system of the frog. III. Clones stemming from individual blastomeres of the 128–, 256–, and 512–cell stages. *J. Neurosci* 3:1019–38; 1983.

————. "Origins of the Nervous System in Amphibians." In *Neuronal Development,* edited by N. C. Spitzer, New York: Plenum Press, 1982.

James, P. D. *Death in Holy Orders.* New York: Ballantine Books, 2001.

Jazwinski, S. M. Longevity, genes, and aging. *Science* 273:54–59; 1996.

Johnson, J., J. Canning, T. Kaneko, J. K. Pru, and J. L. Tilly. Germline stem cells and follicular renewal in the postnatal mammalian ovary. *Nature* 428, 145–50; 2004.

Johnson, M. H. "Three Types of Cell Interaction Regulate the Generation of Cell Diversity in the Mouse Blastocyst." In *The Cell in Contact: Adhesions and Junctions Are Morphogenetic Determinants,* edited by G. M. Edelman and J-P. Thiery, 27–48. New York: Wiley, 1985.

Jones, P. A., and D. Takai. The role of DNA methylation in mammalian epigenetics. *Science* 293:1068–70; 2001.

Joyce, G. F. The Rise and Fall of the RNA World. *New Biologist* 3:399–407; 1991. Reprinted in *Origins of Life: The Central Concepts,* edited by D. W. Deamer and G. R. Fleischaker, 391–99. Boston: Jones and Bartlett, 1994.

Joyce, G. F., and L. E. Orgel. "Prospects for Understanding the Origin of the RNA World." In *The RNA World: The Nature of Modern RNA Suggests a Prebiotic RNA World,* edited by R. F. Gesteland and J. F. Atkins, 1–25. Cold Spring Harbor: Cold Spring Harbor Laboratory Press, 1993.

Kaeberlein, M., K. T. Kirkland, S. Fields, B. K. Kennedy. Sir2-independent life span extension by calorie restriction in yeast. *PloS* 2; Sept. 2004 (doi: 10.1371/journal.pbio.0020296).

Kamshilov, M. M. *Evolution of the Biosphere.* Translated by M. Brodskaya. Moscow: Mir, 1976.

Kanatsu-Shinohara, M., N. Ogonuki, T. Iwano, J. Lee, Y. Kazuki, K. Inoue, H. Miki, et al. Genetic and epigenetic properties of mouse male germline stem cells during long-term culture. *Develop.* 132:4155–63; 2005 (doi: 10.1242/dev.02004).

Kannisto, V. Frailty and survival. *Genus* 47:101–18; 1991.

Kannisto, V., J. Lauritsen, A. R. Thatcher, and J. W. Vaupel. Reductions in mortality at advanced ages: Several decades of evidence from 27 countries. *Population Review* 20:793–810; 1994.

Kass, L. The case for mortality. *Am. Scholar* 52:173–91; 1983.

Kass, L. R., and J. Q. Wilson. *The Ethics of Human Cloning.* Washington, DC: AEI Press, 1998.

Kauffman, S. A. *The Origins of Order: Self-Organization and Selection in Evolution.* New York: Oxford University Press, 1993.

Kauffman, S. *At Home in the Universe: The Search for the Laws of Self-Organization and Complexity.* New York: Oxford University Press, 1995.

Keller, E. F. *Making Sense of Life: Explaining Biological Development with Models, Metaphors, and Machines.* Cambridge: Harvard University Press, 2002.

Keller, L. Changing the world. *Nature* 425:769–70; 2003.

Kim, M., H-B. Moon, and G. J. Spangrude. Major age-related changes of mouse hematopoietic stem/progenitor cells. *Annals New York Academy Sciences* 996: 195–208; 2003.

Kimmel, C. B., and R. M. Warga. Tissue-specific cell lineages originate in the gastrula of the zebra fish. *Science* 231:365–68; 1986.

Kimmins, S., and P. Sassone-Corsi. Chromatin remodeling and epigenetic features of germ cells. *Nature* 434:583–89; 2005 (doi: 10.1038/nature03368).

Kipling, D., T. Davis, E. L. Ostler, and R. G. A. Faragher. What can progeroid syndromes tell us about human aging? *Science* 305:1426–31; 2004.

Kirkwood, T. *Time of Our Lives: The Science of Human Aging.* Oxford: Oxford University Press, 1999.

Kirkwood, T. B. Comparative life spans of species: Why do species have the life spans they do? *Am. J. Clin. Nutr.* 55:1191S–95S (abstract); 1992.

Kirkwood, T. B., and S. N. Austad, "Why do we age?" *Nature* 408:233–38; 2000.

Kirkwood T. B., and T. Cremer. Cytogerontology since 1881: A reappraisal of August Weismann and a review of modern progress. *Hum. Genet.* 60:101–21; 1982.

Kishi, S., J. Uchiyama, A. M. Baughman, T. Goto, M. C. Lin, and S. B Tsai. The zebra fish as a vertebrate model of functional aging and very gradual senescence. *Exp. Gerontol* 38:777–86; 2003.

Klapper, W., K. Kuhne, K. K. Singh, K. Heidorn, R. Parwaresch, and G. Krupp. Longevity of Lobsters is linked to ubiquitous telomerase expression. *FEBS Lett.* 439:143–46; 1998.

Klapper, W., K. Heidorn, K. Kuhne, R. Parwaresch, and G. Krupp. Telomerase activity in 'immortal' fish. *FEBS Lett.* 434:409–12; 1998 (PMID: 9742964).

Klarsfeld, A., and F. Revah. *The Biology of Death: Origins of Mortality.* Translated by Lydia Brady. Ithaca: Cornell University Press, 2004.

Klein, S. L. The first cleavage furrow demarcates the dorsal-ventral axis in *Xenopus* embryos. *Develop. Biol.* 120:299–304; 1987.

Knoll, A. H., and S. B. Carroll. Early animal evolution: Emerging views from comparative biology and geology. *Science* 284:2129–37; 1999.

Kolata, G. *Clone: The Road to Dolly and the Path Ahead.* New York: William Morrow, 1998.

Krajick, K. Methuselahs in our midst. *Science* 302:768–69; Oct. 31, 2003.

Labosky, P. A., D. P. Barlow, and B. L. Hogan. Mouse embryonic germ (EG) cell lines transmission through the germline and differences in the methylation imprint of insulin-like growth factor 2 receptor (Igf2r) gene compared with embryonic stem (ES) cell lines. *Develop.* 120:3197–204; 1994.

Lahdenper, M., V. Lummaa, S. Helle, M. Tremblay, and A. F. Russell. Fitness benefits of prolonged post-reproductive lifespan in women. *Nature* 428:178–81; 2004 (doi:10.1038/nature02367).

Laird, A. K. Dynamics of relative growth. *Growth* 29:249–63; 1965.

Laird, A. K., S. A. Tyler, and A. D. Barton. Dynamics of normal growth. *Growth* 29:233–48; 1965.

Lajtha, L. G. Stem cell concepts. *Differentiation* 14:23–43; 1979.

Laland, K. N., J. Odling-Smee, and M. W. Feldman. Causing a commotion: Niche construction. *Nature* 429:609; June 10, 2004.

Landy, H. J., and L. G. Keith. The vanishing twin: A review. *Human Reproduction Update* 4:177–83; 1998.

Lansing, A. I. "Biological and Cellular Problems of Ageing: I. General Physiology." In *Cowdry's Problems of Ageing*, 3rd ed., edited by A. I. Lansing, 14–19. Baltimore: Williams and Wilkins, 1952.

Lapo, A. V. *Traces of Bygone Biospheres.* Translated by V. Purto. Moscow: Mir, 1982.

Lau, L. L., and L. M. Spain. Altered aging-related thymic involution in T cell receptor transgenic, MHC-deficient, and CD4–deficient mice. *Mech. Ageing Dev.* 114:101–21; 2000 (doi:10.1016/S0047–6374(00)00091–9).

Leblond, C. P., and B. E. Walker. Renewal of cell populations. *Physiol. Rev.* 36:255–79; 1956.

LeDourin, N.M. *The Neural Crest.* Cambridge: Cambridge University Press, 1982.

Leon, D. *Willful Behaviour.* London: Arrow, 2003.

Leslie, M. Ageless no more. *Sci. Aging Knowl. Environ.* nf9; Feb. 2, 2005 (doi: 10.1126/sagckc.2005.5.nf9).

———. If telomeres thrive, you will survive. *Sci. Aging Knowl. Environ.* 2004, no. 18, nf47; May 5, 2004 (doi: 10.1126/sageke.2004.18.nf47).

Lewin, R. *Complexity: Life at the Edge of Chaos.* New York: Macmillan, 1992.

Li, L., H-C. Hsu, G. E. William, C. R. Stockard, K-J. Ho, P. Lott, P-A. Yang, H-G. Zhang, and J. D. Mountz. Cellular mechanism of thymic involution. *Scandinavian J. Immunol.* 57:410–22; 2003 (doi:10.1046/j.1365–3083.2003.01206.x).

Lincoln, G. A., and R. V. Short. Seasonal breeding: Nature's contraceptive. *Recent Prog. Hormone Res.* 36:1–43; 1980.

Ljungquist, B., S. Berg, J. Lanke, G. E. McClearn, and N. L. Pedersen. The effect of genetic factors for longevity: A comparison of identical and fraternal twins in the Swedish Twin Registry. *J. Gerontol. Medical Sciences* 53A:M441–6; 1998.

Lloberas, J., and A. Celada. Mini-Review: Effect of aging on macrophage function. *Exp. Gerontol.* 37:1325–31; 2002 (doi:10.1016/S0531–5565(02)00125–0).

Lolle, S. J., J. I. Victor, J. M. Young, and R. E. Pruitt. Genome-wide non-mendelian inheritance of extra-genomic information in Arabidopsis 505; 2005 (doi: 10.1038/nature03380).

Longo, V. D., and C. E. Finch. Evolutionary medicine: From dwarf model systems to healthy centenarians? *Science* 299:1342–46; 2003.

Lorenz, E. N. *The Essence of Chaos.* Seattle: University of Washington Press, 1993.

Lovelock, J. *The Ages of Gaia.* New York: Bantam Books, 1990.

———. "The Gaia Hypothesis." In *Gaia in Action,* edited by P. Bunyard, 13–31. Edinburgh: Flores, 1996.

Lovibond, D. No way to grieve. *The Spectator* 295, no 9181:16; July 24, 2004.

Lu, T., Y. Pan, S. Y. Kao, C. Li, I. Kohane, J. Chan, and B. A. Yankner. Gene regulation and DNA damage in the ageing human brain. *Nature* 429:883–91; 2004.

Macfarlane Burnet, F., Sir. *Self and Not-self: Cellular Immunology: Book One.* London: Cambridge University Press, 1969.

Macklon, N. S., J. P. M. Geraedts, and B. C. J. M. Fauser. Conception to ongoing pregnancy: The 'black box' of early pregnancy loss. *Human Reproduction Update* 8:333–43; 2002.

Mahowald, A. P. "Genetic Analysis of Oogenesis and Determination." In *Time, Space, and Pattern in Embryonic Development,* edited by W. R. Jeffery and R. A. Raff, 349–63. New York: Alan R. Liss, 1983.

———. The germ plasm of *Drosophila:* A model system for the study of embryonic determination. *Am. Zool.* 17:551–63; 1977.

Mahowald, A. P., C. D. Allis, K. M. Karrer, E. M. Underwood, and G. L. Waring. "Germ Plasm and Pole Cells of *Drosophila*." In *Determinants of Spatial Organization.* Thirty-Seventh Symposium of the Society for Developmental Biology, edited by S . Subtelny, 127–46. New York: Alan R. Liss, 1979.

Mair, W., P. Goymer, S. D. Pletcher, and L. Partridge. Demography of dietary restriction and death in *Drosophila. Science* 301:1731–33; 2003.

Malthus, T. R. *An Essay on the Principle of Population and a Summary View of the Principle of Population.* Edited with an introduction by Antony Flew. London: Penguin, 1970. (Originally published in 1798.)

Manton, K. G., and E. Stallard. Longevity in the United States: Age and sex-specific evidence on lifespan limits from mortality patterns 1960–1990. *J. Gerontol.* 51A:B362–75; 1996.

Margulis, L. and D. Sagano. *What Is Sex.* New York: Simon and Schuster Editions, 1997.

Margulis, L., J. O. Corliss, M. Melkonian, and D. J. Chapman. *Handbook of Protoctista.* Boston: Jones and Bartlett, 1990.

Margulis, L., K. V. Schwartz, and M. Dolan. *The Illustrated Five Kingdoms: A Guide to the Diversity of Life on Earth.* Illustrated by K. Delisle and C. Lyons. New York: HarperCollins, 1994.

Marshman, E., C. Booth, and C. S. Potten. The intestinal epithelial stem cell. *BioEssays* 24:91–98; 2002.

Martin, G. R. Isolation of a pluripotent cell line from early mouse embryos cultured in medium conditioned by teratocarcinoma stem cells. *Proc. Natl. Acad. Sci. USA* 78:7634–38; 1981.

Martin, K., T. B. L. Kirkwood, and C. S. Potten. Age changes in stem cells of murine small intestinal crypts. *Exp. Cell Res.* 241:316–23; 1998.

Mathe, G. Immunity aging. I. The chronic perturbation of the thymus acute involution at puberty? Or the participation of the lymphoid organs and cells in fatal physiologic decline? *Biomed. Pharmacother* 51:49–57 (abstract); 1997.

Mattson, M. P., and G. Van Zant. "Preface." In *Stem Cells: A Cellular Fountain of Youth. Advances in Cell Aging and Gerontology*, edited by M. P. Mattson and G. Van Zant, vol. 9, 2002.

Maturana, H. R., and F. J. Varela. *The Tree of Knowledge: The Biological Roots of Human Understanding.* Rev. ed. Translated by Robert Paolucci. Boston: Shambhala, 1998.

Maynard Smith, J. Review lectures on senescence. 1. The causes of ageing. *Proc. Royal Soc. London, Series B. Biological Sciences* 157:115–27; 1962.

Maynard Smith, J., and E. Szathmáry. *The Origins of Life: From the Birth of Life to the Origin of Language* Oxford: Oxford University Press, 1999.

Mayr, E. *The Growth of Biological Thought: Diversity, Evolution, and Inheritance.* Cambridge: Belknap Press, 1982.

McAuley, P. J. *Pasquales's Angel.* London: Millennium, 1999.

McCarter, R. J. M. "Effects of Exercise and Dietary Restriction on Energy Metabolism and Longevity." In *Modulation of Aging Processes by Dietary Restriction*, edited by Byung Pal Yu, 157–74. Boca Raton: CRC Press, 1994.

McDowell, A. J. Age at menarche, United States, 1973. Rockville: National Center for Health Statistics, series 11, number 133, 1–29, Jan. 23, 2003.

McManners, J. *Death and the Enlightenment: Changing Attitudes to Death among Christians and Unbelievers in Eighteenth-Century France.* New York: Oxford University Press, 1981.

Mechelli, A., J. T. Crinion, U. Noppeney, J. O'Doherty, J. Ashburner, R. S. Frackowiak, and C. J. Price. Structural plasticity in the bilingual brain. *Nature* 431: 757; Oct. 14, 2004.

Medawar, P. B., and J. S. Medawar. *Aristotle to Zoos: A Philosophical Dictionary of Biology.* Cambridge: Harvard University Press, 1983.

Medvedev, Z. A. An attempt at a rational classification of theories of ageing. *Biol. Rev.* 65:375–98; 1990.

Mestel, R. People in U.S. Living Longer. *Los Angeles Times*, (March 1, 2005): A12.

Mezey, É., S. Key, G. Vogelsang, I. Szalayova, G. D. Lange, and B. Crain. Transplanted bone marrow generates new neurons in human brains. *Proc. Natl. Acad. Sci. USA* 100:1364–69; 2003.

Michalakis, Y., and D. Roze. Epistasis in RNA viruses. *Science* 306:1492–93; 2004.

Migliaccio, E., M. Giorgio, S. Mele, G. Pelicci, P. Reboldi, P. P. Pandolfi, L. Lanfrancone, and P.G. Pelicci. The p66^{shc} adaptor protein controls oxidative stress response and life span in mammals. *Nature* 402:309–13; 1999.

Ministry of Health, Labour, and Welfare. http://www.mhlw.go.jp/english/database/db-hw/lifetb03/1.html (2003).

Mintz, B., and K. Illmensee. Normal genetically mosaic mice produced from malignant teratocarcinoma cells. *Proc. Natl. Acad. Sci. USA* 72:3585–89; 1975.

Mitman, G., M. Murphy, and C. Sellers. "Introduction: A Cloud over History." In *Landscapes of Exposure: Knowledge and Illness in Modern Environments*, edited by G. Mitman, M. Murphy, and C. Sellers, 1–17. *Osiris* 19; 2004.

Mojzsis, S. J., G, Arrhenius, K. D. McKeegan, T. M. Harrison, A. P. Nutman, and C. R. Friend. Evidence for life on Earth before 3,800 million years ago. *Nature* 384:55–59; 1996.

Moore, K. L., and T. V. N. Persaud. *The Developing Human: Clinically Oriented Embryology.* 7th ed. Philadelphia: W. B. Saunders, 2002.

Morrison, S. J., P. M. White, C. Zock, and D. J. Anderson. Prospective identification, isolation by flow cytometry, and in vivo self-renewal of multipotent mammalian neural crest stem cells. *Cell* 96:737–49; 1999.

Moyers, B. Blind faith. *In These Times* (Feb. 28, 2005):22–23.

Müller-Sieburg, C. E., R. H. Cho, M. Thoman, B. Adkins, and H. B. Sieburg. Deterministic regulation of hematopoietic stem cell self-renewal and differentiation. *Blood* 100:1302–9; 2002.

Nagel, T. "Death." In *Mortal Questions*, 1–10. Canto edition. Cambridge: Cambridge University Press, 1992.

Narbonne, G.M. Modular construction of early Ediacaran complex life forms. *Science* 305:1141–44; 2004.

National Center for Health Statistics: http://www.cdc.gov/nchs (accessed Feb. 28, 2005).

National Park Service. "Great Basin." U.S. Department of the Interior, http://www.nps.gov/grba/.

Nelson, J. F. "Neuroendocrine Involvement in the Retardation of Aging by Dietary Restriction: A Hypothesis." In *Modulation of Aging Processes by Dietary Restriction*, edited by Byung Pal Yu, 37–55. Boca Raton: CRC Press, 1994.

Nemoto, S., and T. Finkel. Ageing and the mystery at Arles. *Nature* 429:149–52; May 13, 2004.

Nesse, R. M., and G. C. Williams. *Evolution and Healing: The New Science of Darwinian Medicine.* London: Phoenix, 1996.

Nicholls, H. One of a kind. *Nature* 429:498–500; June 3, 2004.

Nieuwkoop, P. D., and L. A. Sutasurya. *Primordial Germ Cells in the Chordates: Embryogenesis and Phylogenesis.* Cambridge: Cambridge University Press, 1979.

Noodén, L. D. "Whole Plant Senescence." In *Senescence and Aging in Plants*, edited by L. D. Noodén and A. C. Leopold, 392–439. San Diego: Academic Press, 1988.

Noto T., and H. Endoh. A "chimera" theory on the origin of dicyemid mesozoans: Evolution driven by frequent lateral gene transfer from host to parasite. *Biosystems* 73:73–83; 2004.

Noyes, B. Experimental studies on the life-history of a rotifer reproducing parthenogenetically (*Proales decipiens*). *J. Exp. Zool.* 35:225–55; 1922. (Arranged and quoted in Pearl, 1924.)

Nybo Andersen, A-M., J. Wohlfahrt, P. Christens, J. Olsen, and M. Melbye. Maternal age and fetal loss: Population based register linkage study. *Brit. Med. J.* 320:1708–12; 2000.

Nyström, T. Aging in bacteria. *Current Opinion in Microbiology* 5:596–601; 2002 (doi:10.1016/S1369–5274(02)00367–3).

Ohinata, Y., B. Payer, D. O'Carroll, K. Ancelin, Y. Ono, et al. Blimp1 is a critical determinant of the germ cell lineage in mice. *Nature* 436:207–13; 2005.

Olby, R. C. Horticulture: The font for the baptism of genetics. *Nature Reviews: Genetics* 1:65–70; 2000.

Olshansky, S. J., and B. A. Carnes, *The Quest for Immortality: Science at the Frontiers of Aging.* New York: Norton, 2001.

Olshansky, S. J., B. A. Carnes, and C. Cassel. In search of Methuselah: Estimating the upper limits to human longevity. *Science* 250:634–40; 1990.

Oriard, M. "Introduction." In *The Abysmal Brute*, by Jack London, v–xiv. Lincoln: University of Nebraska Press, 2000.

Ortman, C. L., K. A. Dittmar, P. L. Witte, and P. T. Le. Molecular characterization of the mouse involuted thymus: Aberrations in expression of transcription regulators in thymocyte and epithelial compartments. *Int. Immunol.* 14:813–22; 2002.

Orwig, K. E., B-Y. Ryu, M. R. Avarbock, and R. L. Brinster. Male germ-line stem cell potential is predicted by morphology of cells in neonatal rat testes, *Proc. Natl. Acad. Sci. USA* 99:11706–11; 2002 (doi: 10.1073/pnas.182412099).

Orwig, K. E., T. Shinohara, M. R. Avarbock, and R. L. Brinster. Functional analysis of stem cells in the adult rat testis. *Biol. Reprod.* 66:944–49; 2002.

Oxford English Dictionary. 2nd ed. and 2006. Online. http://dictionary.oed.com/

Pace, N. R. A molecular view of microbial diversity and the biosphere. *Science* 276:734–40; 1997.

Pardee, K., J. Reinking, and H. Krause. Nuclear hormone receptors, metabolism, and aging. *Science* 306:1446–47; 2004.

Pawelec, G., M. Adibzadeh, R. Solana, and I. Beckman. The T cell in the ageing individual. *Mech. Ageing Dev.* 93:35–45 (abstract); 1997.

Pawelec, G., Y. Barnett, R. Forsey, D. Frasca, A. Globerson, J. McLeod, C. Caruso, et al. T cells and aging. *Front. Biosci.* 1, no. 7:d1056–183 (abstract); 2002.

Pawelec, G., E. Remarque, Y. Barnett, and R. Solana. T cells and aging. *Front. Biosci.* 3:d59–99; 1998.

Pear, R. Social Security understates future life span, critics say. *New York Times* (Dec. 31, 2004):A23.

Pearl, R. *The Biology of Death.* Monographs on Experimental Biology. Edited by J. Loeb, T. H. Morgan, and W. J. V Osterhout. Philadelphia: J. B. Lippincott, 1922.

———. *Studies in Human Biology.* Baltimore: Williams and Wilkins, 1924.

———. *The Rate of Living: Being an Account of some Experimental Studies on the Biology of Life Duration.* New York: Alfred A. Knopf, 1928.

———. *Man the Animal.* Bloomington: Principia Press, 1946.

Pearl, R., and R. DeWitt Pearl. *The Ancestry of the Long-Lived.* New York: Arno Press, 1979. (Originally published in 1934.)

Pearl, R., S. L. Parker, and B. M. Gonzalez. Experimental studies on the duration of life. VII. The Mendelian inheritance of duration of life in crosses of wild type and quintuple stocks of *Drosophila melanogaster. Am. Naturalist* 57:153–92; 1923.

Pearson, K. *The Chances of Death and Other Studies in Evolution.* In two volumes. Vol. 1. New York: Edward Arnold, 1897.

Pelicci, P. G. Perspective Series: Do tumor-suppressive mechanisms contribute to organism aging by inducing stem cell senescence? *J. Clin. Invest.* 113:4–7; 2004.

Penin, X. C. Berge, and M. Baylac. Ontogenetic study of the skull in modern humans and the common chimpanzees: Neotenic hypothesis reconsidered with a tridimensional procrustes analysis. *Am. J. Phys. Anthropol.* 118:50–62; 2002 (doi: 10.1002/ajpa.10044).

Perez, G. I., A. M. Trbovich, R. G. Gosden, and J. L. Tilly. Reproductive biology: Mitochondria and the death of oocytes. *Nature* 403:500–1; 2000.

Perls, T. T., L. Alpert, and R. C. Fretts. Middle-aged mothers live longer. *Nature* 389:133; 1997.

Perls, T. T., and H. Silver. *Living to 100: Lessons in Living to Your Maximum Potential at Any Age.* New York: Basic Books, 2000.

Pikarsky, E., R. M. Porat, I. Stein, R. Abramovitch, S. Amit, S. Kasem, E. Gutkovich-Pyest, S. Urieli-Shoval, E. Galun, and Y. Ben-Neriah. NF-kB functions as a tumour promoter in inflammation-associated cancer. *Nature* 431:461–67; 2004.

Pimentel, D. Changing genes to feed the world. A review of N. Fedoroff and N. M. Brown, *Mendel in the Kitchen: A Scientist's View of Genetically Modified Foods. Science* 306:815; 2004.

Plas, E., P. Berger, M. Hermann, and H. Pfluger. Effects of aging on male fertility? *Exp. Gerontol.* 35:543–51; 2000.

Potten, C. S. A comparison of cell replacement in bone marrow, testis and three regions of surface epithelium. *Biochem. Biophys. Acta* 560:281–99; 1979.

Potten, C. S., and M. Loeffler. Stem cells: Attributes, cycles, spirals, pitfalls and uncertainties: Lessons for and from the crypt. *Development* 110:1001–20; 1990.

Prigogine, I., and I. Stengers. *Order Out of Chaos: Man's New Dialogue with Nature.* New York: Bantam, 1984.

Provine, W. B. "Genetics." In *The Evolutionary Synthesis: Perspectives on the Unification of Biology*, edited by E. Mayr and W. B. Provine, 51–58. Cambridge: Harvard University Press, 1980.

———. *The Origins of Theoretical Population Genetics.* Chicago: Chicago University Press, 1971.

Raff, M. Adult stem cell plasticity: Fact or artifact? *Annu. Rev. Cell Dev. Biol.* 19:1–22; 2003.

RajBhandary, U. L., and D. Söll. "Transfer RNA in Its Fourth Decade." In *tRNA: Structure, Biosynthesis and Function*, edited by D. Söll and U. L. RajBhandary, 1–4. Washington, DC: ASM Press, 1995.

Rao, M. S., and M. P. Mattson. Stem cells and aging: Expanding the possibilities. *Mech. Ageing Dev.* 122:713–34; 2001.

Raup, D. M. *Extinction: Bad Genes or Bad Luck?* New York: Norton, 1991.

Renaulta, V., L-E. Thornell, G. Butler-Browne, and V. Moulya. Human skeletal muscle satellite cells: Aging, oxidative stress and the mitotic clock. *Exp. Gerontol.* 37:1229–36; 2002 (doi:10.1016/S0531–5565(02)00129–8).

Reya, T., and H. Clevers. Wnt signaling in stem cells and cancer. *Nature* 434:843–50; 2005.

Reynaud, G. Capacités reproductrices et descendance de poulets ayant subi un transfert de cellules germinales primordiales Durant la vie embryonnaire. *Wilhelm Roux' Arch. Dev. Biol.* 179:85–110; 1976.

——— Transfer de cellules germinales primordiales de dindon à l'embryon de poulet par injection intravasculaire. *J. Embryol. exp. Morphol.* 21:485–507; 1969.

Rhim, J. A., E. P. Sandgren, J. L. Degen, R. D. Palmiter, and R. L. Brinster. Replacement of diseased mouse liver by hepatic cell transplantation. *Science* 263:1149–52; 1994.

Rice, A. C., A. Khaldi, H. B. Harvey, N. J. Salman, F. White, H. Fillmore, and M. R. Bullock. Proliferation and neuronal differentiation of mitotically active cells following traumatic brain injury. *Exp. Neurol.* 183:406–17; 2003.

Rice, S. H. The analysis of ontogenetic trajectories: When a change in size or shape is not heterochrony. *Proc. Natl. Acad. Sci. USA* 94:907–12; 1997.

Rieger, R. M., P. Ladurner, and B. Hobmayer. A clue to the origin of bilateria? *Science* 307:353–54; Jan. 21, 2005.

Rivera, M. C., and J. A. Lake. The ring of life provides evidence for a genome fusion origin of eukaryotes. *Nature* 431:152–55; 2004, Comment *Nature* 431:134–37; 2004.

Robert, J. S. Model systems in stem cell biology. *BioEssays* 26:1005–12; 2004.

Robin, C, K. Ottersbach, M. de Bruijn, X. Ma, K. van der Horn, and E. Dzierzak, Developmental origins of hematopoietic stem cells. *Oncol. Res.* 13:315–21; 2003.

Roche, A. F. preparer. Executive summary of workshop to consider secular trends and possible pooling of data in relation to the revision of the NCHS growth charts. Hyattsville: Division of Health Examination Statistics, National Center for Health Statistics, 1–47, 1995. http://www.cdc.gov/.

Roenneberg, T., T. Kuehnle, P. P. Pramstaller, J. Ricken, M. Havel, A. Guth, and M. Merrow. A marker for the end of adolescence. *Curr. Biol.* 14:R1038–39; Dec. 29, 2004.

Rose, M. *Darwin's Spectre: Evolutionary Biology in the Modern World.* Princeton: Princeton University Press, 1998.

Roth, G. S., J. A. Mattison, M. A. Ottinger, M. E. Chachich, M. A. Lane, and D. K. Ingram. Aging in Rhesus monkeys: Relevance to human health interventions. *Science* 305:1423–26; 2004.

Rowe, J. W., and R. L. Kahn. *Successful Aging.* New York: Pantheon, 1998.

Rudnicki, M. A., and M. W. McBurney. "Cell Culture Methods and Induction of Differentiation of Embryonal Carcinoma Cell Lines." In *Teratocarcinomas and Embryonic Stem Cells: A Practical Approach*, edited by E. J. Roberston, 19–49. Oxford: IRL Press, 1987.

Rudolph, K. L., S. Chang, H. W. Lee, M. Blasco, G. J. Gottlieb, C. Greider, and R. A. DePinho. Longevity, stress response and cancer in aging telomerase-deficient mice. *Cell* 96:701–12; 1999.

Ruse, M. *The Darwinian Revolution: Science Red in Tooth and Claw.* 2nd ed. Chicago: University of Chicago Press, 1999.

Ryu, B. Y., K. E. Orwig, M. R. Avarbock, and R. L. Brinster. Stem cell and niche development in the postnatal rat testis. *Develop. Biol.* 263:253–63; 2003.

Safi, R., W. Bertrand, O. Marchand, M. Duffraisse, A. de Luze, J. M. Vanacker, M. Maraninchi, A. Margotat, V. Demeneix, and V. Laudet. The axolotl (*Ambystoma mexicanum*), a neotenic amphibian, expresses functional thyroid hormone receptors. *Endocrinol.* 145:760–72. Epub 2003 Oct. 23. 2004.

Sanjuán, R., A. Moya, and S. F. Elena. The contribution of epistasis to the architecture of fitness in an RNA virus. *Proc. Natl. Acad. Sci. USA* 101:15376–79; 2004.

Sano, Y., K. Terada, Y. Takshashi, and A. P. Nutman. Origin of life from apatite dating? *Nature* 400:127; 1999.

Sapolsky, R. M., L. C. Krey, and B. S. McEwen, The neuroendocrinology of stress and aging: The glucocorticoid cascade hypothesis. *Endocrine Rev.* 7:284–301; 1986.

Sauer, M. V., R. J. Paulson, and R. A. Lobo. Oocyte donation to women of advanced reproductive age: Pregnancy results and obstetrical outcomes in patients 45 years and older. *Human Reproduction* 11:2540–43;1996.

Saunders, J. W., Jr., and J. F. Fallon. "Cell death in morphogenesis." *Major Problems in Developmental Biology.* Twenty-fifth Symposium of the Society for Developmental Biology, edited by M. Locke, 289–314. New York: Academic Press, 1966.

Schiffmann, Y. Self-organization in biology and development. *Prog. Biophys. Molec. Biol.* 68:145–205; 1997.

Schlegel, M., J. Lom, A. Stechmann, D. Bernhard, D. Leipe, I. Dyková, and M. L. Sogin. Phylogenetic analysis of complete small subunit ribosomal RNA coding regions of *Myxidium lieberkuehni*: Evidence that Myxozoa are Metazoa and related to the Bilateria. *Arch. Protistenke* 147:1–9; 1996.

Schofield, R. The relationship between the spleen colony-forming cell and the haemopoietic stem cell. A hypothesis. *Blood Cells* 4:7; 1978.

Schopf, J. W. *Cradle of Life: The Discovery of Earth's Earliest Fossils.* Princeton: Princeton University Press, 1999.

Schultz, E. Satellite cell proliferative compartments in growing skeletal muscle. *Develop. Biol.* 175:84–94; 1996.

Schwann, T. *Mikroscopische Untersuchungen über die Uebereinstimmung in der Structur und dem Wachsthum der Thiere und Planzen.* Translated by H. Smith, 1847, in *Sydenham Soc.* 12. Reprinted in part in *Great Experiments in Biology*, edited by M. L. Gabriel and S. Fogel. Englewood Cliffs: Prentice-Hall, 1955. (Originally published in 1839.)

Seipel, K., N. Yanze, and V. Schmid. The germ line and somatic stem cell gene *Cniwi* in the jellyfish *Podocoryne carnea*. *Int. J. Develop. Biol.* 48:1–7; 2004.

Senapathy, P. *Independent Birth of Organisms: A New Theory that Distinct Organisms Arose Independently from the Primordial Pond, Showing that Evolutionary Theories Are Fundamentally Incorrect.* Madison: Genome Press, 1994.

Shakespeare, W. Reproduced from the English Verse Drama Full-Text Database Copyright © [1994 1995] Chadwyck-Healey Ltd. http://collections.chadwyck.com/

Shamblott, M. J., J. Axelman, S. Wang, E. M. Gugg, J. W. Littlefield, P. J. Donovan, P. D. Blumenthal, G. R. Huggins, and J. D. Gearhart. Derivation of pluripotent stem cells from cultured human primordial germ cells. *Proc. Natl. Acad. Sci. USA* 95:13726–31; 1998.

Shapiro, M. D., M. E. Marks, C. L. Peichel, B. K. Blackman, K. S. Nereng, B. Jónsson, D. Schulter, and D. M. Kingsley. Genetic and developmental basis of evolutionary pelvic reduction in threespine sticklebacks. *Nature* 428:717–23; 2004

Sharpless N. E., and R. A. DePinho. Perspective Series: Telomeres, stem cells, senescence, and cancer. *J. Clin. Invest.* 113:160–168; 2004 (doi:10.1172/JCI200420761).

Shaw, B. *One Million Tomorrows*. New York: Ace, 1970.

Sherins, R. J. Are semen quality and male fertility changing? *N. Engl. J Med.* 332:327–28; 1995.

Shimokawa, I., and H. Yoshikazu, "Effect of Dietary Restriction on Pathological Processes." *Modulation of Aging Processes by Dietary Restriction*, edited by Byung Pal Yu, 247–65. Boca Raton: CRC Press, 1994.

Shostak, S. *Becoming Immortal: Combining Cloning and Stem-cell Therapy*. Albany: SUNY Press, 2002.

———. Budding in *Hydra viridis*. *J. Exp. Zool.* 169:430–46; 1968.

———. *Death of Life: The Legacy of Molecular Biology*. London:·Macmillan, 1999.

———. (Re)defining stem cells. *BioEssays* 28:301–08; 2006.

———. *The Evolution of Sameness and Difference: Reflections on the Human Genome Project*. Amsterdam: Harwood Academic Press, 2001.

Shostak, S., J. W. Bisbee, C. Askin, and R. V. Tammariello. Budding in *Hydra viridis*. *J. Exp. Zool.* 169:423–30; 1968.

Shostak, Seth. *Sharing the Universe: Perspectives on Extraterrestrial Life*. Berkeley: Berkeley Hills Books, 1998.

Shubin, N. H., and R. D. Dahn. Lost and found. *Nature* 428:703–4; April 15, 2004.

Siddal, M. E., D. S. Martin, D. Bridge, S. S. Desser, and D. K. Cone. The demise of a phylum of protists: Phylogeny of Myxozoa and other parasitic Cnidaria. *J. Paraasitol.* 81:961–67; 1995.

Simpson, J. L., J. L. Mills, L. B. Holmes, C. L. Ober, J. Aarons, L. Jovanovic, and R. H. Knopp. Low fetal loss rates after ultrasound-proved viability in early pregnancy. *JAMA* 258:2555–57; 1987.

Skurk, T., and H. Hauner. Obesity and impaired fibrinolysis: Role of adipose production of plasminogen activator inhibitor-1. *Int. J. Obesity*, advance online publication Aug. 31, 2004 (doi: 10.1038/sj.ijo.0802778).

Slack, J. M. W. *From Egg to Embryo: Determinative Events in Early Development*. Cambridge: Cambridge University Press, 1983.

———. Stem cells in epithelial tissues. *Science* 287:1431–33; 2000.

Smith, A. Want a long life? Better talk to the old man. *Sidney Morning Herald* (March 2, 2004).

Smith, A. G. Embryo-derived stem cells: Of mice and men. *Annu. Rev. Cell Develop. Biol.* 17:435–62; 2001.

Smith K. R., G. P. Mineau, and L. L. Bean. Fertility and post-reproductive longevity. *Soc. Biol.* 49:185–205; 2003.

Smothers, J. F., C. D. von Dohlen, L. H. Smith, Jr., and R. D. Spall. Molecular evidence that the myxozoans protists are metazoans. *Science* 265:1719–21; 1994.

Social Security Trustees' Annual Report. The Social Security Administration, http://www.ssa.gov/OACT/TR/TR04/index.html. (Updated March 23, 2004.)

Sozou, P. D., and R. M. Seymour. To age or not to age. *Proceedings: Biological Sciences* 271:457–63; March 7, 2004 (doi: 10.1098/rspb.2003.2614).

Spalding, K., R. Bhardwaj, B. Buchholz, H. Druid, and J. Frisén. Retrospective birth dating of cells in humans. *Cell* 122:133–43; 2005.

Spradling, A. C. Stem cells: More like a man. *Nature* 428:133–34; 2004 (doi:10.1038/428133b).

Stearns, S. C. *The Evolution of Life Histories.* Oxford: Oxford University Press, 1992.

Stedman's Medical Dictionary. 25th ed. Baltimore: Williams and Wilkins, 1990.

Stem Cells Book #01 0680. Sponsored by the National Institutes of Health, http://stemcells.nih.gov/info/basics.asp.

Stevens, L. C. Embryonic potency of embryoid bodies derived from a transplantable testicular teratoma of the mouse. *Develop. Biol.* 2:285–97; 1960.

———— "Teratogenesis and Spontaneous Parthenogenesis in Mice." In *The Developmental Biology of Reproduction.* Thirty-Third Symposium of the Society for Developmental Biology, edited by L. Markert and J. Papaconstantinou, 93–106. New York: Academic Press, 1975.

Stewart, E. J., R. Madden, G. Paul, and F. Taddei. Aging and death in an organism that reproduces by morphologically symmetric division. *PLoS Biol.* e45, published online Feb. 1, 2005 (doi: 10.1371/journal.pbio.0030045).

Strome, S., and W. B. Wood. Generation of asymmetry and segregation of germ-line granules in early *C. elegans* embryos. *Cell* 35:15–25; 1983.

Sudo, K., H. Ema, Y. Morita, and H. Nakuchi. Age-associated characteristics of murine hematopoietic stem cells. *J. Exp. Med.* 192:1273–80; 2000.

Sulston, J. E., E. Schierenberg, J. G. White, and H. N. Thomson. The embryonic cell lineage of the nematode *Caenorhabditis elegans. Develop. Biol.* 100:64–119; 1983.

Surani, M. A. Stem cells: How to make eggs and sperm. *Nature* 427:106–7; Jan. 8, 2004.

Susser, M. "Environment and Biology in Aging: Some Epidemiological Notions." In *Aging: Biology and Behavior*, edited by James L. McGaugh and Sara B. Kiesler, 77–96. New York: Academic Press, 1981.

Szabo, P., K. Zhao, I. Kirman, J. Le Maoult, R. Dyall, W. Cruikshank, and M. E. Weksler. Maturation of B cell precursors is impaired in thymic-deprived nude and old mice. *J. Immunol.* 161:2248–53; 1998.

Takeoka Y., S. Y. Chen, H. Yago, R. Boyd, S. Suehiro, L. D. Shultz, A. A. Ansari, and M. E. Gershwin. The murine thymic microenvironment: Changes with age. *Int. Arch Allergy Immunol.* 111:5–12; 1996.

Tarkowski, A. K. "Induced Parthenogenesis in the Mouse." In *The Developmental Biology of Reproduction*, Twenty-Third Symposium of the Society for Developmental Biology, edited by C. L. Markert and J. Papaconstantinou, 107–29. Orlando: Academic Press, 1975.

Tatar, M., A. Bartke, and A. Antebi. The endocrine regulation of aging by insulin-like signals. *Science* 299:1346–51; 2003.

Thoman, M. L. Early steps in T cell development are affected by aging. *Cellular Immunol.* 178:117–23; 1997 (doi:10.1006/cimm. 1997.1133).

Tiedemann, H. "Substances with morphogenetic activity in differentiation of vertebrates." In *The Biochemistry of Animal Development.* Vol. 3, *Molecular Aspects of Animal Development,* edited by R. Weber, 257–92. Orlando: Academic Press, 1975.

Timiras, P. S. "Demographic Comparative and Differential Aging." *Physiological Basis of Aging and Geriatrics,* 2nd ed., edited by P. S. Timiras, 7–21. Boca Raton: CRC Press, 1994.

Tipler, E. F. *The Physics of Immortality: Modern Cosmology, God and the Resurrection of the Dead.* New York: Doubleday, 1994.

Todes, D. *Darwin without Malthus: The Struggle for Existence in Russian Evolutionary Thought.* New York: Oxford University Press, 1989.

Toichi, E., K. Hanada, T. Hosokawa, K. Higuchi, M. Hosokawa, S. Imamura, and M. Hosono. Age-related decline in humoral immunity caused by the selective loss of TH cells and decline in cellular immunity caused by the impaired migration of inflammatory cells without a loss of TDTH cells in SAMP1 mice. *Mech. Ageing Dev.* 99:199–217; 1997 (doi:10.1016/S0047–6374(97)00100–0).

Tong, S., S. Meagher, and B. Vollenhoven. Dizygotic twin survival in early pregnancy. *Nature* 416:142; 2002.

Torella, D., M. Rota, D. Nurzynska, E. Musso, A. Monsen, I. Shiraishi, E. Zias, et al. Cardiac stem cell and myocyte aging, heart failure, and insulin-like growth factor-1 overexpression. *Circ. Res.* 94:411–13; 2004 (doi:10.1161/01.RES.0000117306. 10142.50).

Toyooka, Y., N. Tsunekawa, R. Akasu, and T. Noce. Embryonic stem cells can form germ cells *in vitro. Proc. Natl. Acad. Sci. USA* 100:11457–62; 2003.

Trifunovic, A., A. Wredenberg, M. Malkenberg, J. N. Spelbrink, A. T. Rovio, C. E. Bruder, M. Bohlooly-Y, et al. Premature ageing in mice expressing defective mitochondrial DNA polymerase. *Nature* 429:417–23; 2004.

Troyansky, D.G. *Old Age in the Old Regime: Image and Experience in Eighteenth-Century France.* Ithaca: Cornell University Press, 1989.

Tumbar, T., G. Guasch, V. Greco, C. Blanpain, W.E. Lowry, M. Rendl, and E. Fuchs. Defining the epithelial stem cell niche in skin. Originally E published *Science Express* on Dec. 11, 2003; *Science* 303:359–63; Jan. 16, 2004 (doi: 10.1126/science.1092436).

Turksen, K. Revisiting the bulge. *Develop. Cell* 2:454–56; 2004.

Turner, S. J. Marc Tartar on the components of aging. *George Street Journal.* (online) file:///research/Frailty/frailtyMarcTatar04–13–04_files/redir.html (2002).

Valentine, J.W. *On the Origin of Phyla.* Chicago: University of Chicago Press, 2004.

Valenzuela, R. K., S. N. Forbes, P. Keim, and P. M. Service. Quantitative trait loci affecting life span in replicated populations of *Drosophila melanogaster*. II. Response to selection, *Genetics* 168:313–24; 2004 (doi: 10.1534/genetics.103.023291).

van der Koory, D., and S. Weiss. Why stem cells? *Science* 287:1439–41; Feb. 25, 2000.

Vaughan, H. *Silex Scintillans* (1650). Reproduced from the English Verse Drama Full-Text Database Copyright © [1994–1995] Chadwyck-Healey Ltd. http://collections.chadwyck.com/

Vaupel, J. W., J. R. Carey, and C. Kaare. It's never too late. *Science* 301:1679–80; Sept. 19, 2003.

Vaupel, J. W., and B. Jeune, "The emergence and proliferation of centenarians." In *Exceptional Longevity: From Prehistory to the Present*, edited by B. Jeune and J. W. Vaupel, 109–116. Monographs on Population Aging, 2. Odense: Odense University Press, 1995.

Vaupel, J. W., K. G. Manton, and E. Stallard. The impact of heterogeneity in individual frailty on the dynamics of mortality. *Demography* 16:439–54; 1979.

Ventura, S. J., J. C. Abma, W. D. Mosher, and S. Henshaw. Revised pregnancy rates, 1990–97, and new rates for 1988–99, United States. *National Vital Statistics Reports*, vol. 52, no. 7. Hyattsville, Md.: National Center for Health Statistics, 2004.

Vossbrinck, C. R., J. V. Maddox, S. Friedman, B. A. Debrunner-Vossbrinck, and C. R. Woese. Ribosomal RNA sequence suggests microsporidia are extremely ancient eukaryotes. *Nature* 326:411–14; 1987.

Vreeland, R. H., W. D. Rosenzweig, and D. W. Powers. Isolation of a 250 million-year-old halotolerant bacterium from a primary salt crystal. *Nature* 407:897–900; 2000.

Vyff, S. "Confessions of a Proselytizing Immortalist." In *The Scientific Conquest of Death: Essays on Infinite Lifespans*, edited by Immortality Institute, 223–32. Buenos Aires: LibrosEn Red; 2004.

Wadman, M. Scientists cry foul as Elsevier axes paper on cancer mortality. *Nature* 429:687; June 17, 2004 (doi:10.1038/429687a).

Wagers, A. J., R. I. Sherwood, J .L. Christensen, and I. L. Weissman. Little evidence for developmental plasticity of adult hematopoietic stem cells. *Science* 297:2256–59. Epub 2002.

Wall Street Journal Online (March 10, 2005).

Wallace, A. R. to A. Newton. 1887. In *The Life of Charles Darwin*, by F. Darwin, London: Senate, 1995. (Originally published in 1902.)

Washburn, S. L. "Longevity in Primates." In *Aging: Biology and Behavior*, edited by James L. McGaugh and Sara B. Kiesler, 11–29. New York: Academic Press, 1981.

Watanabe, K., and S. Osawa. "tRNA Sequences and Variations in the Genetic Code." In *tRNA: Structure, Biosynthesis and Function*, edited by D. Söll and U. L. RajBhandary, 225–50. Washington, DC: ASM Press, 1995.

Weismann, A. *The Germ-plasm: A Theory of Heredity.* Translated by W. Newton Parker and Harriet Rionnfeldt. New York: Charles Scribner's Sons, 1912. (Originally published in German in 1886; first English translation 1893).

———. *Ueber die Dauer des Lebens.* Jena: Gustav Fischer Verlag, 1882.

Weissman, I. L. Translating stem and progenitor cell biology to the clinic: Barriers and opportunities. *Science* 287:1442–46; 2000.

Weissman, I. L., D. J Anderson, and F. Gage. Stem and progenitor cells: Origins, phenotypes, lineage commitments, and transdifferentiation. *Annu. Rev. Cell Develop. Biol.* 17:387–403; 2001.

Williams, G. C. Natural selection, the costs of reproduction and a refinement of Lack's principle. *Am. Nat.* 20:687–90; 1966.

———. Pleiotropy, natural selection and the evolution of senescence. *Evolution* 11:398–411; 1957.

Williamson, D. I. *Larvae and Evolution: Toward a New Zoology.* New York: Chapman and Hall, 1992.

Willier, B. H. Experimentally produced sterile gonads and the problem of the origin of germ cells in the chick embryo. *Anat. Rec.* 70 (Suppl. 1):89–112; 1937.

Wilmoth, J. R., and H. Lundström. Extreme longevity in five countries. *Eu. J. Pop.* 12:63–93; 1996.

Wilson, E. B. *The Cell in Development and Inheritance.* New York: Macmillan. 1896.

———. *The Cell in Development and Inheritance,* Third ed. with corrections. New York: Macmillan, 1925.

———. Experimental studies on germinal localization. I. The germ-region in the egg of Dentalium. II. Experiments on the cleavage-mosaic in Patella and Dentalium. *J. Exp. Zool.* 1:1–72, 197–268; 1904a, b.

Winitsky, S. O., T. V. Gopal, S. Hassanzadeh, H. Takahashi, D. Gryder, M. A. Rogawski, K. Takeda, Z. U. Yu, Y. H. Xu, and N. D. Epstein. Adult murine skeletal muscle contains cells that can differentiate into beating cardiomyocytes in vitro. *PloS Biol.* 3(4):e87; 2005 (doi: 10.1371/journal.pbio.0030087).

Winter, E., J. Wang, M. J. Davies, and R. Norman. Early pregnancy loss following assisted reproductive technology treatment. *Human Reproduction* 17:3220–23; 2002.

Wise, P. M., K. M. Krajnak, M. L. Kashon. Menopause: The aging of multiple pacemakers. *Science* 273:67–70; 1996.

Wobus, A. M., and K. R. Boheler. Embryonic Stem Cells: Prospects for Developmental Biology and Cell Therapy. *Physiol. Rev.* 85:635–78; 2005

Woese, C. R. Bacterial evolution. *Microbiol. Rev.* 51:221–27; 1987.

———. "The Use of Ribosomal RNA in Reconstructing Evolutionary Relationships among Bacteria." In *Evolution at the Molecular Level,* edited by R. K. Selander, A. G. Clark, and T. S. Whittam, 1–24. Sunderland, MA: Sinauer Associates, 1991.

————. The universal ancestor. *Proc. Natl. Acad. Sci. USA* 95:6854–59; 1998.

Woese, C. R., and G. E. Fox. Phylogenetic structure of the prokaryotic domain: The primary kingdoms. *Proc. Natl. Acad. Sci. USA* 74:5088–90; 1977.

Woese, C. R., and R. R. Pace. "Probing RNA structure, function, and history by comparative analysis." In *The RNA World: The Nature of Modern RNA Suggests a Prebiotic RNA World*, edited by R. F. Gesteland and J. F. Atkins, 91–117. Cold Spring Harbor: Cold Spring Harbor Laboratory Press, 1993.

Woese, C. R., O. Kandler, and M. L. Wheelis. Towards a natural system of organisms: Proposal for the domains Archaea, Bacteria, and Eucarya. *Proc. Natl. Acad. Sci. USA* 87:4576–79; 1990.

Woldringh, C. L. Is *Escherichia coli* getting old? *Bioessay* 27:771–74; 2005.

Wolfram, S. *A New Kind of Science.* Champaign, Ill.: Wolfram Media, 2002.

World Health Organization Mortality Database: http://www3.who.ch/whosis/mort/table1.cfm?path=mort,mort_table1&language=english (2004).

Wowk, B. "Medical Time Travel." In *The Scientific Conquest of Death: Essays on Infinite Lifespans*, edited by Immortality Institute, 135–49. Buenos Aires: LibrosEn Red; 2004.

Ye, P., and D. E. Kirschner. Reevaluation of T cell receptor excision circles as a measure of human recent thymic emigrants. *J. Immunol.* 168:4968–79 (abstract); 2002.

Yorke, J., and T-Y. Li. Period three implies chaos. *American Mathematical Monthly* 82:985–92; 1975.

You, L., R. S. Cox III, R. Weiss, and F. H. Arnold. Programmed population control by cell-cell communication and regulated killing. *Nature* 428:868–71; 2004.

Young R. Biological Limits to the Human Life Span: Semantics: e-mail to Gerontology Research Group mailing list, March 20, 2004, http://lists.ucla.edu/cgi-bin/mailman/listinfo/grg.

————. Evolutionary theories of aging and mortality deceleration: e-mail to Gerontology Research Group mailing list, April 16, 2005, http://lists.ucla.edu/cgi-bin/mailman/listinfo/grg.

Yu, L., A. Alva, H. Su, P. Dutt, E. Freundt, S. Welsh, E. H. Baehrecke, and M. J. Leonardo. Regulation of an *ATG7–beclin 1* program of autophagic cell death by caspase-8. *Science* 304:1500–2; 2004.

Zhao, M., S. Momma, K. Delfani, M. Carlén, R. M. Cassidy, C. B. Johansson, H. Brismar, O. Shupliakov, J. Friesén, and A. M. Janson. Evidence for neurogenesis in the adult mammalian substantia nigra. *Proc. Natl. Acad. Sci. USA* 100:7925–30; 2003.

Zinsser, H. *Rats, Lice and History: A Chronicle of Pestilence and Plagues.* New York: Black Dog & Leventhal, 1934.

Index

abortion, 10, 32, 36, 69, 178n100
adolescence, 37, 49, 51–52, 53, 62, 70, 86; preadolescent, 70, 115; post-adolescent, 97; post-pubescent, 84, 115
adulthood, 48, 70, 104, 136, 151; midlife crisis, 49; old adult, 37, 49, 51–52, 128; senescent 37, 43, 49, 51, 62, 70, 128; size of adult, 81, 115–116, young adult, 35, 37, 43–44, 49, 51–52, 70, 128, 131
age-dependent deterioration, 25, 37, 38, 71, 98, 130, 137–139, 139; Alzheimer's, 72; aortic valve calcification, 72; arteriosclerosis; 48, 120; atherosclerosis; 72, 137; autoimmune disease, 25, 27, 137; behavioral deficits, 72; cataracts, 25, 82, 120; collagen accumulation, 139; dementias, 38, 72, 137; fibrinolysis, 71; graying, 120; Huntington's, 16, 38; hypersensitivity, delayed-type, 138; dyslipidaemia, 71; hypertension, 71; hypo-cellular bone marrow, 139, immuno-surveillance breakdown, 137; impotence, 62; interstitial fibrosis, 72; macular degeneration, 25; mental, 38, 48 metabolic syndrome, 71, 72; neurodegeneration, 16; obesity, 71, 186n29; osteoarthritis, 137; osteoporosis, 120, 137; presbycusis, 25; presbyopia, 25; retinal degeneration, 16, 25, 139; unresponsiveness, 137
anti-aging medicine, 32, 130, 135; Centre on Ageing, 24; MacArthur Foundation, 16, 175n27
aging phenotype, 20; aging genes, 16, 20, 39, 120; collateral inheritance, 22; direct lineal inheritance, 22–23; gerontic genes, 20; Hutchinson-Gilford syndrome, 22; premature aging, 120; progeria, 22; Werner's syndrome, 22
algae, 60, 65, 143, 167, 181n18; euglenids, 167
altricial development, 135–136, 144
amphibians, 81, 122, 134, 135, 143; axolotl (*Ambystoma*), 134; mud puppy, 135, 136; *Necturus maculosus*, 135; *Proteus anguinus*, 135; salamander, 134; tadpole, 90
animals (metazoans), 86, 92, 101, 108, 112, 115, 127, 134, 136, 142, 163; eusocial, 162, 170, 171–172, 195n17; evolution of, 77, 92, 134; multicellular, 64, 76, 169; with negligible senescence, 106, 115, 120; sexual life cycle in, 86. *See also* invertebrates, vertebrates
antagonistic serial pleiotropy, 17
antigens, 124–125, 130, 137, 138, 139; embryonic, 125
apoptosis, 16, 121, 138, 139–140, 166
Archaea, 60, 91, 162, 164
autophagia, 16, 166
autopoiesis, 78, 93

baby boomers, 154
Bacteria, 9, 38, 60, 75, 91, 137, 138, 162, 164–165; aging and death in, 118, 165, 195n7; Bacillus/Lactobacillus, 165; biofilms, 164; *Escherichia coli,* 61, 138; *Mycobacterium tuberculosis*, 138; *Pseudomonas aeruginosa*, 138; ribosomes, 166; sexual pili, 164; *Streptococcus pneumoniae*, 138